BIOLOGICAL STRUCTURE AND FUNCTION 6

BLOOD VESSELS

BIOLOGICAL STRUCTURE AND FUNCTION

EDITORS

R. J. HARRISON
Professor of Anatomy
University of Cambridge

R. M. H. McMINN
Professor of Anatomy
Royal College of Surgeons of England

J. E. TREHERNE
Reader in Invertebrate Physiology
University of Cambridge

1. *Biology of Bone*: N. M. HANCOX
2. *The Macrophage*: B. VERNON-ROBERTS
3. *The Integument*: R. J. C. SPEARMAN
4. *The Pituitary Gland*: R. L. HOLMES AND J. N. BALL
5. *The Mammalian Kidney*: D. B. MOFFAT
6. *Blood Vessels*: W. J. CLIFF

THE BLOOD VESSELS

W. J. A. CLIFF

Senior Fellow, Department of Experimental Pathology
John Curtin School of Medical Research
Australian National University

CAMBRIDGE UNIVERSITY PRESS

CAMBRIDGE
LONDON · NEW YORK · MELBOURNE

Published by the Syndics of the Cambridge University Press
The Pitt Building, Trumpington Street, Cambridge CB2 1RP
Bentley House, 200 Euston Road, London NW1 2DB
32 East 57th Street, New York, NY 10022, USA
296 Beaconsfield Parade, Middle Park, Melbourne 3206, Australia

© Cambridge University Press 1976

First published 1976

Printed in Great Britain
at the
University Printing House, Cambridge
(Euan Phillips, University Printer)

Library of Congress cataloguing in publication data

Cliff, Walter John, 1932–
Blood vessels

(Biological structure and function; 6)

Bibliography: p.

Includes index.

1. Blood vessels. I. Title. II. Series.
[DNLM: 1. Blood vessels. 2. Vascular diseases.
W1 BI759L v. 6/WG500 C637b]
QP106.C54   596′.01′16   74–31789
ISBN 0 521 20753 3

TO ASH

CONTENTS

Preface	page	vii
Acknowledgments		1
1 General features of circulatory systems		1
2 The vessel lumen		8
Cross-sectional area		8
Flow velocity		10
Capillaries		14
Collecting vessels		17
Arterial–venous relationships		23
Portal and pulmonary venous systems		23
Sinusoidal vessels		26
3 The endothelium		31
Cell junctions		31
Cytoplasmic processes		37
Cell surfaces and cell shape		39
Organelles and inclusions		43
Pores, permeability and fenestrae		53
Nucleus and mitosis		61
Aging		66
4 The extra-endothelial cells of blood vessel walls		68
Pericytes		68
Vascular smooth muscle cells		72
The fibroblast		91
Phagocytic cells		93
5 The extracellular components of blood vessel walls		97
The basement membrane system		97
Collagen		102
Elastica		110
6 Ancillary structures of blood vessel walls		125
Vasa vasorum		125
Lymphatics of blood vessel walls		130
Innervation of blood vessel walls		133

Contents

7 Vessels as functional units	*page*	141
The arteries		141
Arterio-venous anastomoses (AVAs)		145
Microvascular networks		147
Post-capillary venules of lymphoid tissue		155
The veins		155
References		157
Author index		193
Subject index		203

PREFACE

The compulsion towards excellence in science has resulted in the development of a great degree of specialization. Many scientists limit their research and reading to what, at times, seem exceedingly narrow fields. A worker may devote years to studying one particular cell type or one protein. We frequently find ourselves in the paradoxical situation of being surrounded by mountains of knowledge which might just as well not exist because we are confined within the narrow clefts of our own specialities.

I have attempted to write a book which will help to broaden a little the narrow compass which tends to enclose that group of people who are involved with or interested in cardiovascular research. To many people it must appear presumptuous on my part that I should even attempt to write a monograph on what is now such a vast subject as the blood vessels. No single person could be qualified in all fields to collate, review and criticize the vast amount of information that has accumulated concerning these structures. This book is therefore written very much from the viewpoint of a morphologist. It is, however, my conviction that a study of the anatomy of blood vessels which proceeds from the macroscopic to the microscopic and finally to the ultrastructural, as revealed by electron microscopy, demonstrates that their functions are implicit in their structures and that their structures are entirely subservient to their functions. The relation of normal structure and function is Physiology, while the relation of abnormal function and altered structure is Pathology (see figure). These themes are pursued in considering the major diseases that affect blood vessels and Western Man, such as atherosclerosis, hypertension and aging.

I have, wherever possible, related the morphology to the histochemistry, biochemistry, biophysics and bio-engineering of blood vessels. The book commences by sketching the development of circulatory systems within the invertebrates and the emergence of the double circulation that exists in mammals and birds. The manner in which vessels distribute blood to and within the tissues and return it from the tissues is considered. Studies on vessel lumen diameters, branching, rheology and vessel wall physics are all taken into consideration. Various vascular counter-current systems that exist in different sites and animals are detailed.

The individual components that go to form blood vessel walls are then considered separately. These are the cellular elements, namely endo-

Preface

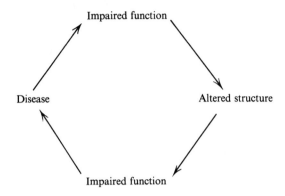

thelium, pericytes, smooth muscle cells, fibroblasts and macrophages and the extracellular components collagen and elastic tissue, the ground substance and basement membranes and, finally, what are essentially ancillary structures, the vasa vasorum, nerves and lymphatics. The permeability of vessels to water, solutes and cells is considered in both physiological and pathological conditions.

In conclusion, after what has been largely a reductionist approach, certain integrated functions of blood vessel walls are considered. The main features of elastic and muscular arteries, arterioles, arterio-venous anastomoses, the microvascular networks, the post-capillary venules and the veins are described and discussed.

ACKNOWLEDGMENTS

It is a pleasure to acknowledge my gratitude to Professor F. C. Courtice, until recently my Head of Department, without whose benevolence and encouragement this project would neither have been undertaken nor completed. I wish to thank Professor H. D. Attwood who read the manuscript most carefully and made many useful suggestions, most of which were incorporated in the final drafting. I am greatly indebted to the generosity of the donors of a number of illustrations that grace the pages of this monograph. They are acknowledged by name in the relevant legends.

Acknowledgment is also due to my technical officer, Mr P. R. Allen, for the unstinting assistance he has given me in all aspects of my work. Mr S. R. Butterworth and Mr R. Westen helped me prepare the figures for publication. To Mrs M. M. Lee goes my heartfelt thanks and admiration for all the work she has done in preparing the typescript in a highly skilled and meticulous manner.

1

GENERAL FEATURES OF CIRCULATORY SYSTEMS

It is to Claude Bernard (1813–78) that we owe the concept of the constancy of the internal environment of living creatures. With the evolution of multicellular organisms certain bio-engineering adaptations have occurred which clearly operate to maintain this internal constant state. Circulatory systems are foremost in these adaptations. In the simpler forms of animals, such as the Protozoa and sponges, an adequate balance can be maintained between the interior of the organisms and the external environment by direct exchange through diffusion and pinocytosis (Prosser, 1950). The coelenterates have developed an efficient means of maintaining a continuous passage of their external watery environment through their interiors. They thus ensure that their internal endodermal cells are continuously being bathed by fresh oxygen and food-bearing water.

With the development of the triploblastic organisms some form of circulatory system within the new mesodermal layer that separates the ecto- and endodermal components rapidly becomes essential. The Platyhelminthes, or flatworms, are an exception to this, but their case is most useful to examine. These animals possess a highly developed system of branching canals which extends throughout their bodies and which functions as a specialized excretory system. This underlines the fundamental importance of the efficient removal of the waste products of metabolism within the economy of the animal and emphasizes the importance of this function of the circulatory systems that are found in all higher forms of animals.

The circulatory systems of invertebrates are in general considerably simpler than those of vertebrates and commonly are characterized by the presence of a ventral heart, the interposition of the gills or lungs in the venous return circuit to the heart and in most cases by the absence of closed capillary network systems (Prosser, 1950; Nishimaru, 1969; Pentreath and Cottrell, 1970). The haemolymph of invertebrates is distributed through a system of arterial vessels to enter large lacunar spaces which encroach considerably on the coelomic space. The tissues are bathed in slowly percolating pools of fluid that is taken up through openings in the walls of the terminal venous vessels to be returned to the heart. It is

2 General features of circulatory systems

not surprising in such situations, where the functions of blood and tissue fluid are inextricably mingled, that no separate lymphatic system occurs, such as is found in the vertebrates (Yoffey and Courtice, 1970). An exception to this rule is found in the highly active cephalopods, with average aortic blood pressures of the order of 40 and 54.5 mmHg being found in two species (*Biological Handbooks*, 1971). Closed capillary networks and structures interpreted as lymphatic vessels have been identified in their tissues (Nishimaru, 1969).

It should be emphasized that no really clear-cut distinction can be drawn between the invertebrate and vertebrate types of circulatory systems. This is seen very clearly in the lowly members of the Vertebrata, the Myxini or hagfish which have a sinusoidal form of circulation in their tissues requiring accessary caudal, cardinal and portal hearts to pump the blood into the venous return vessels (Johansen and Martin, 1965).

Animals, as opposed to plants and bacteria, depend ultimately on oxidative processes as the source of the energy upon which their metabolism depends. Clearly, in the various highly active and also at times very large animals that occur in the Vertebrata, effective means of providing oxygen and of removing carbon dioxide, the end product of organic oxidative processes, must operate. Indeed, in their excellent review on the comparative aspects of cardiovascular function in vertebrates Johansen and Martin state 'The efficiency of the cardiovascular system depends upon the success with which the respiratory organs are adequately connected with the terminal sites of respiration, the metabolizing body mass.' It can be seen that with increasing evolutionary development within the Vertebrata, leading to the assumption of a terrestrial way of life, to the homeothermic form of body temperature control, to the development of speed and agility, or sheer size and finally to intelligence, there have developed, *pari passu*, increasingly efficient circulatory and respiratory systems. Whilst it cannot be maintained that a cause and effect relationship exists linking each of these major steps with the increasing efficiency of these systems, it can be shown that any reduction in the efficiency of these systems rapidly results in impaired tissue function and, in severe instances, to tissue necrosis. A dramatic example is the very rapid loss of consciousness, or syncope, associated with cerebral hypoxia secondary to decreased blood supply to the brain. Less alarming to the uninitiated, but equally instructive, are the trophic changes that occur in the extremities associated with atherosclerotic narrowing of the supplying arterial vessels which in severe cases can lead to the development of dry gangrene (Illingworth, 1955) as a consequence of insufficient blood supply.

Each improvement in the basic design of the vertebrate circulatory system has been of significance, so that a study of this system, most particularly in relation to the branchial vessels, is of use in helping trace the development of the various groups and species within the Vertebrata (Romer, 1950). The initial and major step is taken in the fish, with the development of a closed circulation which allows for a great reduction in blood volume and, as a corollary, a greatly increased efficiency in its utilization (Johansen and Martin, 1965; Prosser, 1950). A progressive reduction in lacunar and sinusoidal forms of circulation can be traced through the elasmobranchs and teleosts, leading to the general pattern in which blood flow occurs through an arterial distributing system to supply networks of fine nutritive vessels present within the tissues from where it is collected into progressively larger veins to be returned to the heart. Such a closed system involves an inevitable frictional energy loss (Green, 1950) and requires an efficient pumping system to impart energy to the circulating blood. A comparison of blood pressures determined in various invertebrates with open circulation patterns and various types of fish gives an index of this improved efficiency. In fact salmon have blood pressures as high as 75 mmHg in their ventral aortae. Such a closed circulatory system partitions the blood from the tissue fluid and is associated with the development of the lymphatic system that is present in all vertebrates (Yoffey and Courtice, 1970) without which virtually the entire plasma volume and 40% of the plasma proteins would be lost to the extracellular fluid each day (Courtice, Simmonds and Steinbeck, 1951).

Perhaps the biggest design fault in the circulation in fish is the interpolation of the branchial, or gill, circulation directly in series with the systemic circulation so that the blood is distributed through two sets of exchange vessels consecutively with attendant drop in dorsal aortic pressure and the undesirable situation of the heart receiving the oxygen-poor blood returning from the tissues for its own metabolic needs. To overcome this deficiency the coronary supply to the heart evolved as a derivative of the branchial circulation of fish. The coronary vessels are considered to be homologous throughout the vertebrate series (Johansen and Martin, 1965) and the similarity of their pharmacological properties to those of the branchial vessels of fish has been pointed out (Prosser, 1950).

Another major change in the circulatory pattern in vertebrates is the appearance of parallel systemic and respiratory circulations as is found in Amphibia. Here, whilst there is still only one ventricle, the spiral valve of the bulbus cordis of the heart functions at least in part to direct the streams of oxygen-rich blood and oxygen-poor blood into the systemic and pulmonary–cutaneous systems respectively (Simons and Michaelis, 1953). In fact Simons and Michaelis found in one quarter of their experiments a highly selective distribution of blood between these two systems, but as pointed out by Johansen and Martin, the function of the amphibian bulbus cordis and spiral valve is complex and poorly understood.

4 General features of circulatory systems

In Reptilia the progressive isolation of the two parallel circuits becomes more and more complete so that in Crocodilia the complete isolation of the two is functionally achieved by the closure of the incomplete interventricular septum during cardiac systole (Romer, 1950). In both Mammalia and Aves this separation is anatomically complete. The possibility of a separate pulmonary circulation existing was first propounded by Michael Servetus (1511–1553) but its physiological discovery was due to William Harvey (1578–1657). Franklin (1949) has published a most interesting and readable monograph on the history of physiology, which for quite a large part is the history of that great integrative system, the circulation, to which the reader is referred for the historical development of our modern concepts.

The great benefit of the double circulation is that although both the pulmonary and systemic circulations have identical volumes of blood flow in a given time, these are obtained at greatly different arterial blood pressures. In man, for instance, the mean systemic arterial pressure is 90 mmHg whilst the mean pulmonary arterial pressure is only 8 to 19 mmHg (Fowler, Westcott and Scott, 1953). The low pulmonary blood pressure allows the highly specialized thin-walled exchange vessels in the walls of the pulmonary alveoli to be in extremely close contact with the epithelium lining the air spaces of the alveoli (Policard, Collet and Pregermain, 1957) with the minimum of mechanical supporting structures, ensuring extremely efficient exchange of oxygen and carbon dioxide over the minimal diffusion distance. This occurs in the presence of pronounced pulsatile flow being transmitted to the pulmonary capillaries from the pulmonary artery (Karatzas *et al.*, 1970). It would appear that in order to obtain the high volume–low pressure configuration the smoothing effect of narrow calibre, high resistance vessels must be largely absent. In fact, the venous elements of the pulmonary circulation contribute up to 18% of total pulmonary vascular resistance in certain circumstances (Furnival, Linden and Snow, 1970). Any derangement of this high volume–low pressure flow pattern can have very rapid and serious repercussions, as evidenced by the ease with which pulmonary oedema can develop as a result of raised pulmonary venous pressure, most often due to impaired left ventricular function (Hunter, 1956).

The high pressure systemic circulation is maintained by two mechanisms working in concert. These are the work done by the highly muscular left ventricle which supplies energy to the contained blood and the vascular tone of the small arteries and arterioles that supply the necessary resistance to blood flow within the system (Green *et al.*, 1944; Green, 1950; Krogh, 1959; Wiederhielm, 1965; Rodbard, 1971). In fact, by analogy with electrical circuits, there is an Ohm's law of the blood circulation. It is generally advantageous to consider the conductance Q/P of a vascular bed rather than its resistance P/Q, where P is the hydrostatic pressure in dynecm^{-2} and Q is volume rate of flow in mlsec^{-1} (Rodbard, 1971). Krogh (1959) also proposed that there was a distinct capillariomotor mechanism involved in the control of peripheral resistance.

The existence of a high pressure systemic circulation brings with it definite advantages. The high pressure–low volume arterial system means that considerable reserves are available for increasing flow rates in regions of high oxygen demand, as for instance in skeletal muscle associated with exercise and post-exercise hyperaemia, by dilatation of the tonic small arteries and arterioles of the region (Krogh, 1959; Duling and Berne, 1971; Mellander and Lundvall, 1971; Scott and Radawski, 1971). A measure of the reserves available is given by the experimental observations of Krogh (1959) who established that the area of blood channels flowing in 1 cm^3 of muscle increased from a control level of 3 cm^2 to 750 cm^2 with exercise. The active hyperaemia of muscle is considered to be due to direct action by metabolically produced alterations in the local environment upon the vascular smooth muscle cells of small arteries and arterioles. These changes have been claimed to be tissue hyperosmolality by Mellander and Lundvall (1971) whilst Scott and Radawski (1971) discount the importance of this factor, except for a minor degree of influence in coronary vessels, and consider that raised Mg^{2+} levels are of major importance in producing active hyperaemia. Active hyperaemia in the cerebral and pial vessels, which are distinguished by their very high degree of intrinsic control, is produced by raised carbon dioxide levels (Shenkin and Bouzarth, 1970; Raper, Kontos and Patterson, 1971). The marked intrinsic local autoregulatory control of the blood flow of the brain has been contrasted to the external hormonal control of the blood flow of the kidney, the first being entirely dictated by the local metabolic needs of the brain tissues and the second being entirely subservient to the needs of producing large volumes of glomerular filtrate (Thurau, 1971). Mellander and Johansson (1968) consider that local hypoxia, K$^+$ concentration and hyperosmolality all operate to produce active hyperaemia. The hyperaemia occurring in adipose tissue stimulated to mobilize free fatty acids, is very long lasting and appears to be linked to the production of an acid–ether extractable material present in the active adipose tissue (Lewis and Matthews, 1970).

Hyperaemia associated with secretion in salivary glands has both an immediate nerve-mediated vasodilatation (Gautvik, 1970; Karpinski, Barton and Schachter, 1971) and a slower developing kinin-mediated component (Gautvik, 1970). A similar dual mechanism of initial nerve-mediated vasodilatation supported later by a metabolic component has been demonstrated in voluntary muscle hyperaemia (Ellison and Zanchetti, 1971). It was demonstrated that the nervous control was sympathetic cholinergic-mediated and to be under the control of the cerebral cortex in trained cats.

6   General features of circulatory systems

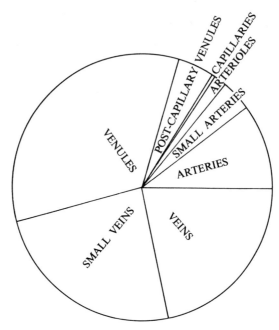

Figure 1. The percentage distribution of blood volume throughout the vascular bed of the bat's wing. (From Wiedeman, 1963. Dimensions of blood vessels from distributing artery to collecting vein. *Circulation Res.*, **12**, 375. Reproduced by permission of The American Heart Association, Inc.)

Having a large volume–low pressure venous system in circuit with the high pressure arterial system means that there are large reserves of blood volume on immediate call to meet the needs of the latter. The venous vessels have very aptly been designated as capacitance vessels. Green (1950) estimated that 66% of the blood volume is present in the venous components of the circulation while Mellander and Johansson (1968) and Wiedeman (1963) gave an estimate of up to 80% of regional blood volumes as venous (figure 1). This ability of the venous side of the circulation to act as a reservoir is highly developed in diving animals (Franklin, 1928) and is associated with the presence of sphincters which constrict the inferior vena cava to reduce venous return to the heart (Walker and Attwood, 1960; Elsner, Hanafee and Hammond, 1971). The volume compliance of veins has been analysed by Moreno *et al.* (1970) and contrasted to the behaviour of latex tubes of similar dimensions. The reaction of venous walls to increasing blood volumes is considerably more complex than is the case with the latex tube models and has both 'bending' and 'stretching' changes occurring simultaneously. Active tone changes in the wall of the superior vena cava have been studied by Guntheroth and Chakmakjian (1971) and are mediated by circulating catecholamines. The richer adrenergic innervation of the veins of birds suggests that by comparison with those of mammals these vessels play a greater part in controlling the regional distribution of blood and the volume of venous return to the heart (Bennett and Malmfors, 1970).

2

THE VESSEL LUMEN

As an undergraduate I spent many hours examining the philosophical question of whether the bung-hole was part of the barrel. This question can also be put in regard to the hole down the centre, or lumen, of a blood vessel. Certainly, without a lumen a blood vessel would cease to exist as a functional component of the body.

CROSS-SECTIONAL AREA

A consideration of the progressive increase in the total cross-sectional area of the blood vessels, produced by the continuing ramification of the vessels as they extend peripherally from the aorta to the exchange networks within the tissues is most instructive. In fact the smallest cross-sectional area and lowest surface to volume ratio are found in the aorta and the greatest total cross-sectional area and highest surface to volume ratio are found in the exchange networks or 'capillaries'. Green (1950) calculated values for the various blood vessels of a 13 kg dog. The aorta had a cross-sectional area of 0.8 cm^2 whilst the 'capillaries' had a total cross-sectional area of 600 cm^2. The number of 'capillaries' arising by the series of successive branchings initiated at the aorta was estimated at 1.2×10^9. Krogh (1959) estimated from injection specimens that there were 1350 'capillaries' mm^{-2} in the gastrocnemius muscle of the horse whilst in smaller mammals there might be up to 4000 mm^{-2}. The contribution in volume of the capillaries to the total volume of the muscle varied from 3.3% (horse) to 10.6% (dog). The surface areas of 1 ml of blood within the muscles of various animals were also calculated and given as 2700 cm^2 (frog), 5600 cm^2 (dog) and 7300 cm^2 (horse). He estimated the total 'capillary' surface area in an average man to be 6300 square metres.

Intaglietta and Zweifach (1971) used an in-vivo technique to map the vascular components of the rabbit's mesentery and arrived at results that appear fairly comparable, with 4.55×10^6 capillaries per cm^3 of tissue, having an exchange area of 330 cm^2 per cm^3 of tissue, whilst the surface area of 1 ml of blood was estimated as 2360 cm^2. Finally the volume of the blood contained in vessels in the tissues was 0.14 ml per cm^3 of tissue.

Whilst these figures vividly illustrate the nature of the microcirculation it must be remembered that the actual blood volume present within these minute exchange vessels is small in absolute amount. Green estimated it at 4.41% of the total blood volume in the dog, and Wiedeman (1963), working on the microvasculature of the bat wing membrane, also stressed the small total volume contribution of the exchange network capillaries (see figure 1).

The continual branching, or arborization, of the blood vessels which leads to the final distribution of the cardiac output throughout all the microvascular exchange networks of the body has been examined mathematically by Groat (1948) using the formula for arteries

$$Q = kD^n,$$

where Q is flow rate, D is the diameter of the vessel adapted to the flow rate and k and n are constants. Groat measured arteries participating in dichotomous branchings with diameters from 3200 μm down to 24 μm. He found the equating value for n in the expression

$$D_a^n = D_b^n + D_c^n + ...,$$

where a is the parent vessel and b, c etc. are daughter vessels. The mean value he obtained for n was 2.6 with range 2.1–3.1. From these results Green calculated that equal branches arising by bifurcation of a parent vessel will have diameters 0.766 of that of the parent vessel.

The extremely high surface to volume ratio that is achieved in the microvascular exchange networks is of great significance in the efficient exchange of metabolites between the blood and the tissues. However it is also responsible for frictional shear effects that supply the major resistance in the circulatory system (Rodbard, 1970). In the conditions of laminar flow that exist within the microcirculation the frictional energy loss is independent of the roughness of the wall and is given by the expression

$$\text{Energy loss} = \frac{128\mu QL}{\pi D^4} \text{ (gcm}^{-2}\text{)},$$

where μ is the viscosity of blood (gseccm^{-2}), Q is the volume rate of flow (cm^{3}sec^{-1}), L is the length (cm) and D is the lumen diameter (cm). It can be seen from this expression that the quotient is extremely sensitive to any alteration in D, or lumen diameter. The viscosity of blood (μ) is not constant but varies with the size of the blood vessel lumen (Dintenfass, 1967). The phenomenon of axial flow of erythrocytes (Goldsmith, 1970) is of importance in reducing the relative overall viscosity of the flowing blood (Green, 1950), and Rosenblum (1970) has shown that increases in plasma viscosity are more effective in prolonging the transit time through the microcirculation than is an increase in the haematocrit.

Even with the decreasing viscosity of blood present in the very small blood vessels there is greater resistance per unit length in each daughter

10 The vessel lumen

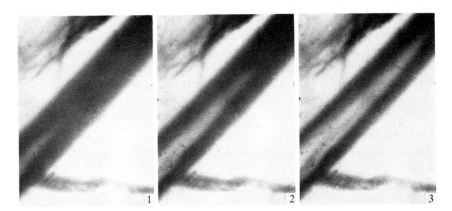

Figure 2. Consecutive frames from a 16 f.p.s. ciné film recording the passage of a bolus injection of dextran–saline in the central artery of the rabbit's ear. The sequence is from left to right and shows the advancing interface which has a parabolic shape produced by the flow velocity gradient. Magnification ×9.

vessel compared with the parent vessel (Groat, 1948). By making hydraulic models of actual microcirculatory beds that were mapped from series of photomicrographs taken *in vivo* Intaglietta and Zweifach (1971) found that the hydraulic hindrance remained uniform from the small arterioles (20 μm diameter) through the microcirculatory networks until the venules were reached when it declined abruptly to 5% of the total value.

FLOW VELOCITY

The velocity of flow within blood vessels is generally considered to have a parabolic profile with the blood in contact with the wall being at, or very near, zero flow and the blood in the centre of the vessel having the highest velocity (figure 2) (Green, 1950; Texon, 1957; Scharfstein, Gutstein and Lewis, 1963; Glagov, 1965). In such a situation the velocity of axial flow is twice the average velocity of blood flow. However, recent direct measurements of blood flow velocity, made using anemometer probes positioned within the lumen of the pulmonary artery of dogs and human beings, showed velocity profiles which were virtually flat from one side of the vessel to the other. In other words there was no evidence of a stationary boundary layer which, with the viscous shear of the flowing blood, would operate to produce a parabolic flow profile (Reuben, Swadling and Lee, 1970).

The presence of a parabolic flow velocity profile has been proposed as an aetiological factor in the development of intimal atherosclerotic lesions related to vessel branches, particularly at bifurcations, when the fast moving axial stream is divided and comes into direct relation with the inside walls of the bifurcated vessel and the abnormal boundary layer conditions so produced cause intimal damage (Texon, 1957; Scharfstein et al., 1963; Glagov, 1965). Caro, Fitz-Gerald and Schroter (1970) have proposed, however, that high shear rates in themselves are beneficial and prevent the development of atheroma. Investigations by Gutstein, Farrell and Schenk (1970) of flow rates and patterns at and downstream from the bifurcation of the aorta into the common iliac arteries have revealed considerable instability in the flow pattern, with changes interpreted by them as indicating breakdown of streamline flow into turbulent flow. What makes their findings so surprising is that this breakdown was at the very low Reynold's number of 100. McDonald (1952a), using high speed cinematography, has also claimed that turbulent flow occurs in the abdominal aorta of rabbits, particularly in relation to its bifurcation into the iliac arteries. However, when read in conjunction with his previous paper (McDonald, 1952b), the interpretation is obscure since he makes no mention of a break-up of the parabolic dye-front profile in these observations which seems to persist through the cardiac cycle. Green (1950) considered that turbulent flow is not likely to occur until the Reynold's number is greater than 2000, while Ellis (1970) gives a figure ten times higher than this. The Reynold's number for a fluid is given by the following expression and is dimensionless:

$$Re = \frac{VD\rho}{\mu},$$

where V is the average flow velocity (cm sec^{-1}), D is the diameter of the lumen (cm), ρ is the density of the fluid (g cm^{-3}) and μ is its viscosity (g sec cm^{-2}). It can be seen from this expression that the Reynold's number will be highest, and so the tendency to develop turbulent flow will be greatest, in vessels that have both large diameters and rapid rates of blood flow. Stehbens (1959) has emphasized that calculations of Reynold's numbers based on the case of uniform straight tubes are inapplicable to the situations found in blood vessels. He has made some very elegant observations with glass tubes that imitate the conditions of curved, branched and bifurcated vessels. He has shown that instability induced by curves, branches and bifurcations can precipitate turbulence at low Reynold's numbers (figure 3). When turbulent flow occurs the orderly laminar or streamline pattern of velocity profile disappears and the fluid then moves as a single mass although individual particles show random velocity and courses (Green, 1950). Such a situation occurring in blood vessels has been linked with the development of atheroma (Glagov, 1965) and with post-stenotic dilatation of arteries (Aars and Solberg, 1971). In

12 *The vessel lumen*

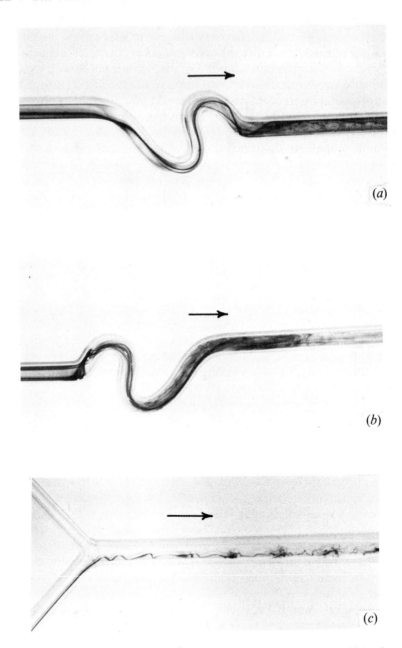

Figure 3. Glass tube models showing the influence that the shape of blood vessels may have in producing turbulent flow at low Reynold's numbers (Re). Arrows show direction of flow. (*a*) Analogue of internal carotid siphon $Re = 567$; (*b*) The same but with reverse flow $Re = 521$; (*c*) Analogue of basilar artery situation $Re = 929$. (From Stehbens, 1959.)

addition, vibrations that are set up in the walls of arteries, as associated with stenosis, may coincide at certain levels with the resonant frequency of the wall. In this situation relatively high strains can be produced in the wall by relatively low forces (Foreman and Hutchison, 1970). Anitschkow (1967) has acknowledged the importance of local haemodynamic factors in the formation of atheromatous plaques.

In any consideration of the shearing forces due to the flowing blood and the wear-and-tear occasioned thereby in the walls of arteries it is important to consider the actual values of the velocity changes and the rates of changes, or accelerations, involved. Values for the peak blood velocity in the dog's aorta have been given as between 30 and 83 $cm\,sec^{-1}$, that in the rabbit's abdominal aorta has been measured as 60 $cm\,sec^{-1}$, whilst that in the human aorta has been calculated by Evans (1918–19) as 40 $cm\,sec^{-1}$. From the illustrations of an on-line Doppler analysis technique for measuring blood flow velocity (Light, 1970) the flow velocity in the human aorta appears to be of the order of 100 $cm\,sec^{-1}$. Taking this last figure for the sake of illustration, a simple calculation shows that the blood is flowing at 3.6 $km\,hr^{-1}$ at its fastest. This is barely the pace of an amble down a country lane! However the flow within the aorta is intermittent. The coronary arterial blood flow is also extremely intermittent, being absent or even retrograde during cardiac systole and attaining its maximum rate only after the completion of isometric relaxation in diastole (Gregg, 1934). Positive aortic flow is present for only one third of the cardiac cycle in mammals and birds, which is in contrast to fish who have positive flow in the ventral aorta for three quarters of the cardiac cycle (Johansen and Martin, 1965). During part of the cardiac cycle the direction of blood flow in the aorta may be retrograde and has been measured by McDonald (1952b) in the rabbit abdominal aorta as attaining a velocity of 27 $cm\,sec^{-1}$. He considered that the back-flow in the aorta, which was extremely variable and could at times be absent, was too great a component to be explicable by the classical concept of momentary aortic valve incompetence before closure. He preferred to look upon the back-flow phenomenon as an expression of arterial run-off occurring from the aorta into the visceral vasculature. This is the 'Windkessel' function of the aorta described by Mellander and Johansson (1968). McDonald (1952a) calculated that the acceleration of the blood within the aorta can be as high as 6000 $cm\,sec^{-2}$, or roughly 6 **g** in modern aerospace jargon, and is obviously a factor worthy of consideration.

The analysis of the physical properties of a series of tapering branched tubes in which pulsatile flow is occurring, in terms of reflection of waves from branches (Arndt, Stegall and Wicke, 1971), Fourier analysis of wall harmonics (Peterson, Jensen and Parnell, 1960), and non-uniform impedance characteristics of the arterial tree (Learoyd and Taylor, 1966), has led Duchacek (1967) to conclude that the analysis of the fluid mechanics of blood flow is as complex as any flow problem can be. He considered the amount of digital computer time required for the solutions of the various problems to be a real drawback to their use in the study of blood flow and suggested that an 'impedance method' similar to that of electrical transmission line theory might be better suited for analysing blood flow problems.

The intermittent pulsatile flow that is present in the arteries is smoothed to a very large degree by the time the blood reaches the microcirculatory exchange networks. My own microscopic observations on the blood flow in systemic vessels in fish, amphibians and mammals have convinced me that in the normal physiological situation there is no pulsatile flow in such vessels. This view is also held by Landis (1926) who made direct measurements of blood pressure variations throughout the microvascular system of the frog mesentery using a microcannula technique and found that the pulse pressure usually flattened out sharply at the beginning of the capillaries proper. Wiederhielm *et al.* (1964) could detect a pulse pressure of 1.1 mmHg in the arterial capillaries of the frog's mesentery but not beyond this level in the microvascular bed. Zweifach (1971*a*) found that the capillary pressures remained constant to within ±2 mmHg in the rat despite considerable variations in the arterial pressure. Williams (1944) observed that the capillary flow in the mouse thyroid was rapid and constant. In the pulmonary circulation, as studied in the dog, the pulmonary capillary bed shows pulsatile flow with an amplitude one half of that in the pulmonary artery and with some evidence of smoothing, in that capillary flow tails off exponentially to the end of cardiac diastole (Karatzas *et al.*, 1970). Krogh (1959) considered that a pulse is transmitted to vessels down to capillary size and is only finally obliterated by phase effects in venules that are draining blood from capillaries of differing lengths.

CAPILLARIES

It is by now far too late in the day to expect to alter the accepted usage of the word 'capillary'. It is derived from the Latin *capillaris* and as such pertains to the hair-like calibre of the small vessels first seen by the early anatomist Malpighi in the frog's lung in 1661. A definition in terms of calibre alone is today unacceptable. I can do no better than quote Wiedeman (1963) who has expressed the modern approach to the definition of blood vessels in the following words: 'It is meaningless to define an artery or arteriole on the basis of its diameter, or to define a capillary in this manner. The assignment of a name to a specific vessel should be determined by its position and function in the vascular system.' The extremely arbitrary nature of the identification of 'capillaries' in the living microcirculation has been stressed also by Grafflin and Bagley (1953).

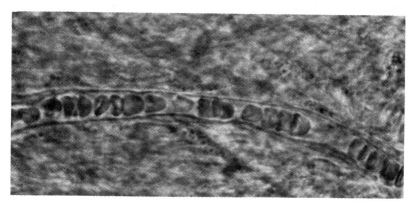

Figure 4. Red blood cells being deformed into bullet or parachute shapes within a very narrow vessel with a fast flow in a rabbit ear chamber. Direction of flow from left to right of figure. Exposure 5 msec, magnification × 850.

The identification, that still occurs in some papers, of 'capillaries' as the smallest vessels identifiable in sections is totally erroneous. The true bottle-neck in the circulatory system is the region of the terminal arteriole and the metarteriole. These vessels are so narrow that erythrocytes are deformed into bullet-like shapes aptly described as 'parachute' cells by Bond, Derrick and Guest (1964) (figure 4). Although these authors describe the narrow fast-flow vessels as 'narrow capillaries' my own observations of this phenomenon have been mainly in vessels I would term metarterioles. Braasch (1971) has drawn attention to the generally overlooked phenomenon in rheology of erythrocyte deformability which is a function of the ratio of the surface area to the volume of the cell. In the usual situation the biconcave mammalian erythrocyte has a considerable amount of 'excess' membrane which makes it readily deformable. The goat is exceptional in having spherical erythrocytes and in this case the minimum pore size that allows passage of these cells corresponds to the actual diameter of the cells (3.5 μm). With other species the minimum pore size is considerably smaller than the diameter of the erythrocytes under test. Goldsmith (1970) has drawn attention to the importance of red cell deformability in producing their migration into the axial stream of flow in a vessel.

I prefer to use functional terms based on in-vivo properties to describe the components of the microcirculation. The so called 'capillaries' of physiology may be identified as the thin-walled exchange vessels of the microcirculation that form anastomosing networks within the tissues which link the narrow, fast-flow, direct and purposeful arterioles and metarterioles to the wider, meandering and branching venules (figure 5). The rate of blood flow within the exchange networks is the slowest in the

16 The vessel lumen

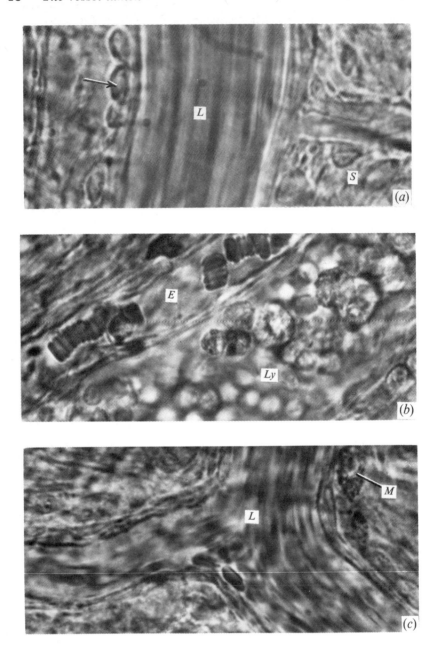

Figure 5. Photomicrographs of blood vessels in rabbit ear chambers taken at 1/20 sec exposure.
(a) An arteriole with lumen (L) flows from top to bottom of figure. It has a single layer of well-defined smooth muscle cells (arrow) in its wall. A metarteriole arising

microcirculation. Once the blood enters these networks it has time to wander, as it were, through tranquil untroubled by-ways. It is most pleasant to watch through the microscope their gentle meanderings as the various blood components go about their business of exchanging metabolites and of tissue surveillance. It is here that all the hurly-burly and hustle and bustle of being collected-up, gathered into veins, pumped around, oxygenated, squirted out and finally squeezed through comes to fruition. Their sojourn in these vessels is, however, all too brief. The transit time for fluorescein from arterioles to venules through the cerebral microcirculation, as determined by Rosenblum (1970) using a ciné technique, is 0.68 sec. This is shorter than the estimated average transit time through capillaries, given by Green (1950) as 1.5 sec. This may well be due to the average capillary length he assumed of 1.00 mm which is higher than values measured by Zweifach (1971a) of 0.25 mm in general, with a maximum of 0.8 mm, in muscle, and than the 0.23 mm average length measured by Wiedeman (1963) in the bat's wing. The average velocity of flow in the capillary net vessels has been measured by Landis (1926) as being 0.5 mm sec^{-1}, whilst Green made an estimate of 0.7 mm sec^{-1} for this factor. Krogh used exactly the same figure as given by Landis of 0.50 mm sec^{-1} in his book *The Anatomy and Physiology of Capillaries* (Krogh, 1959). The short time that the blood actually spends within the exchange vessels of the tissues must be at least in part responsible for the fact that generally not more than 40% of arterial oxygen is utilized by the tissues (Evans, 1949).

COLLECTING VESSELS

There comes a point in the branching and anastomosing of the minute vessels forming the exchange nets in the tissues when they develop the characteristics of collecting vessels. The process of tributaries uniting to form progressively larger collecting, or venous, vessels continues as a somewhat imperfect mirror image of the ramifications of the arterial system. The total cross-sectional area of the venous channels likewise becomes progressively smaller as the venous blood approaches the heart (Green, 1950). The narrowest components of the venous system, the post-

---

from its right side has a smooth muscle cell (S) present near its origin. Magnification ×1300.

(b) A slow-flow exchange vessel (E) of the microcirculation contains red cell rouleaux, platelets and a single granulocyte. A lymphatic (Ly) runs immediately beside this vessel and contains a pleomorphic collection of macrophages (large cells) and leukocytes. Magnification ×850.

(c) The richly-branching and anastomosing venule system has wider lumen (L) and more rapid flow than the type of vessel shown in (b). Macrophages (M) with numerous granular cytoplasmic inclusions are prominently associated with their walls. Magnification ×850.

capillary venules and the very small veins or venules into which they drain (diameters up to ~ 100 μm), have extremely important characteristics that distinguish them quite sharply from the rest of the microcirculation in terms of permeability, reactions in acute inflammation, and functions in lymphoid organs which will be discussed below under the relevant headings.

As veins become larger they generally follow the same course as the arterial vessel supplying the region. Such veins are commonly paired and lie one on each side of the deep arteries, where they are termed *venae commitantes* (Smith, 1967).

The venules and veins return to the heart the blood that has passed through the microcirculatory networks of the tissues. Veins, with only certain exceptions, contain numbers of valves along their lengths (Franklin, 1927-8). The valves are directed so as to permit blood flow to occur only in the direction of the heart (Krogh, 1959) and to prevent regurgitation of venous blood into the tissues (Kampmeier and Birch, 1927). In addition, the valves make the readily distensible thin-walled veins into efficient pumps that actively assist the return of blood to the heart (Franklin, 1928; Krogh, 1959; Folkow *et al.*, 1971). This pumping action is dependent on compression of the vein walls caused by the contraction of voluntary muscles, an effect that has been graphically described as analogous to squeezing a wet sponge. Hence it is found in the limbs that the superficial veins first drain into the deep vein systems to return blood to the heart.

Active venous return in human beings is of great importance in view of the little appreciated fact that a man's heart is at a greater height above his feet than in any other animal save the elephant and the giraffe (Krogh, 1959). Folkow and others (1971) found that the muscle pump system working with efficient venous valves to maintain a low venous pressure was of considerable physiological significance. They found they could increase the effective blood flow in the calf muscles of rhythmically exercising men by 60% by tilting their subjects from the legs horizontal position (supine) to an angle of 60° legs down. This, they proposed, was because by maintaining a low venous pressure the increase of 65-70 mmHg in arterial perfusion pressure produced by tilting the legs down was fully able to operate. Using ^{133}Xe to measure blood flow, they found an enormous venous outflow during muscular contraction and arterial inflow during relaxation. Subjects with varicose veins had lower exercise capabilities than normal subjects, which was considered to be due to incompetent venous valves preventing efficient lowering of the venous pressure.

In addition to special instances of veins within certain sites and organs having no valves (Franklin, 1927-8; Kampmeier and Birch, 1927) it may be said that in general it is the very large and the very small veins that lack valves. It has been proposed that the large valveless intrathoracic and intra-abdominal veins may assume the role of the sinus venosus of lower vertebrates. Batson (1942) has ascribed considerable significance to

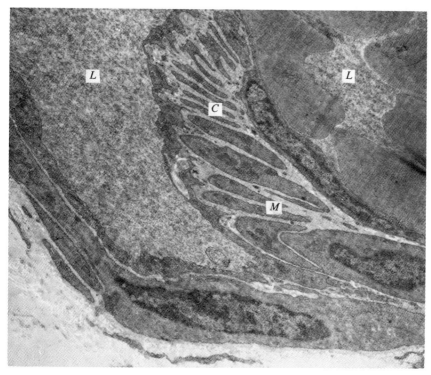

Figure 6. A valve cusp (C) extending into the lumen (L) of a vein in the wing of a bat. The valve cusp contains numerous smooth muscle cell profiles (M) between its two endothelial surfaces. Magnification ×20000.

the vertebral system of veins in the clinical spread of tumour metastases. The vertebral venous system can be demonstrated as a valveless plexiform series of longitudinal venous channels that joins the cranial venous sinuses to those in the pelvis.

The question of the effect of venous contractions on the return of blood to the heart must be considered. In only one group of mammals, the bats, does it appear that the peripheral systemic venous system is contracting phasically all the time (Nicoll and Webb, 1955). In these animals phasic propagated contractions of vein walls can be readily seen by in-vivo microscopy of their wing membranes. In conjunction with the action of numerous muscular valves (figure 6) situated along the lengths of the veins these contractions propel the blood very effectively in a central direction back towards the heart. Whilst the case of the bat wing veins is probably the most graphic, the more generalized nature of venous rhythmic contractility should not be lost sight of (Franklin, 1928). Certainly, considerable work has been done on the mesenteric and portal veins which show spontaneous rhythmic activity in a variety of animals (figure 7) (Sutter,

The vessel lumen

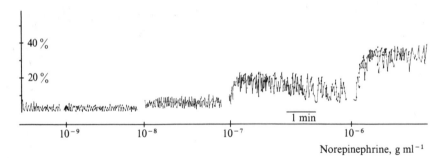

Figure 7. Rhythmic contractions and dose response to norepinephrine of a rat portal vein preparation. Upward displacement indicates shortening expressed as a percentage of initial length.

1965; Voth *et al.*, 1969; Helfer and Jaques, 1970; Ljung, 1970; McConnell and Roddie, 1970). Somlyo, Woo and Somlyo (1965) investigated the spontaneous rhythmic contractions of the nerve-free umbilical vessels and found that the vein was more active than the artery. The presence of striated muscle with typical intercalated discs, such as are seen in cardiac muscle, in the walls of murine pulmonary and large thoracic veins (Karrer, 1960*a*) and the rat's pulmonary veins (Ludatscher, 1968), is evidence that these vessels are very probably rhythmically contractile. As an indication of the contribution that venous contractions may make to general circulatory dynamics Mellander and Johansson (1968) quoted a figure of 25–30% expulsion of regional blood volume associated with constriction of the capacitance vessels.

Where spontaneous rhythmic contraction of veins does occur in mammals it is probably an atavistic property of blood vessels persisting (Hama, 1960; Johansen and Martin, 1965).

ARTERIAL–VENOUS RELATIONSHIPS

The presence of specialized arterial–venous relationships is extremely interesting. Such associations are apparent between the larger blood vessels of the limbs where the arteries and veins are extremely closely related. The close proximity of the two streams of blood, the one entering and the other leaving the extremity, allows for the counter-current exchange of heat to occur between them (Bazett *et al.*, 1948) (figure 8). This method of conserving body heat in which the warm arterial blood coming from the central regions of high body temperature passes almost all its excess heat directly to the cold venous blood returning from the extremity is most graphically illustrated in the ingenious arterial–venous relationships found in the fins of whales (Scholander and Schevill, 1955). Scholander *et al.* (1950) cite cases of Arctic animals that survive tempera-

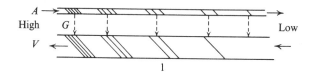
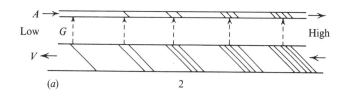

Figure 8. (a) A schematic diagram showing the general relationships between arterial (A) and venous (V) vessels involved in counter-current exchanges. The diagonal lines indicate the relative blood levels of the variable involved (e.g. temperature, pH, oxygen tension). In 1 the arterial venous gradient (G) operates to maintain a high central level and a low peripheral level of the variable. In 2 the reverse situation exists where the arterial–venous gradient (G) operates to maintain a low central level and a high peripheral level of the variable.
(b) A schematic representation of the rete mirabile type of arterial–venous counter-current system where the vessels branch locally to form two sets of numerous parallel, intimately associated channels.

ture differentials of 100 °C between their body temperature and that of their environment. They found that such animals maintained the temperature of their legs at just above 0 °C. The caribou was found to have fats with a 30 °C lower melting point in the distal regions as compared to the proximal regions of their legs. Seagulls can adapt their feet to withstand temperatures of −20° to −40 °C by vasomotor control. The importance of this counter-current mechanism is twofold. Firstly, it conserves the body heat and secondly, it prevents the extremities from being warmed by the arterial blood entering it. It might at first sight seem a rather unhappy situation for an Arctic gull to have such cold feet; however, Scholander *et al.* recount the tale of a seagull that had been kept in a cage at +20 °C for observation while his fellows had been acclimatizing to the −20 °C conditions of the Arctic winter. The captive gull escaped one day from the station and its feet had frozen within one minute of landing on the

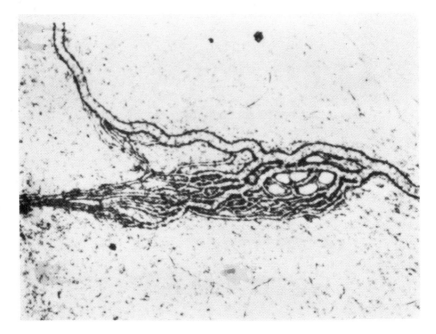

Figure 9. Small arterial vessels enter at the left side of the figure and become intimately related to a leash of tortuous interconnecting venous vessels in the external spermatic fascia of the rat. (Grant and Payling Wright, 1971.)

snow outside. It would seem to me that the feet of the acclimatized birds, being almost at 0 °C, had no tendency to melt the snow but that the initially warm feet of the escaped bird probably did cause melting and in so doing broke down the insulating effect of the air trapped amongst the snow crystals, with disastrous results.

Similar specialized arterial and venous relations are seen in the vascular supply of the mammalian testis where it is essential for the development and viability of the spermatozoa that the temperature of the testis should be several degrees lower than body temperature (Evans, 1949; Romer, 1950). This lower temperature is attained by the location of the testes outside of the abdomen in the scrotal sac and by the presence of a highly developed thermal exchange apparatus in the form of the pampiniform venous plexus that twines and branches in a vine-like manner (hence the name) around the testicular artery and vas deferens (Brash and Jamieson, 1947). Extremely intricate and beautiful systems of arterial–venous relationships considered to have thermoregulatory functions have been found on the spermatic fascia of the rat by Grant and Payling Wright (1971) (figure 9).

PORTAL AND PULMONARY VENOUS SYSTEMS

The function of veins has so far been considered in terms of their role as capacitance elements in the circulation and as the return limb of the circulatory system delivering the blood back to the heart. There are in addition very important instances where veins do not return their contained blood directly to the heart but conduct it to be redistributed throughout a second series of microvascular networks or capillary beds. These are the various portal systems. The most widespread throughout the vertebrates is the hepatic portal system (Romer, 1950) and it is the position of this vein in the porta hepatis from which the term 'portal' originally derives. This vein receives the blood from the territories of the anterior (or superior) mesenteric vein and the splenic vein. The portal vein divides into large branches that enter each lobe of the liver and by progressive branching distributes its blood throughout the entire liver (Michels, 1962). The portal vein supplies 72% of the total hepatic blood flow and the hepatic artery supplies only 28%. The total hepatic blood flow has been estimated in a variety of animals, including man, as being $\sim 25\%$ of the cardiac output (Greenway and Stark, 1971). This agrees with the estimate made by Dickinson and Secker Walker (1970) of the blood volume of the rabbit liver which was 25–30% of the total blood volume. This pool of hepatic blood can be mobilized as shown by its large reduction as a percentage of total blood volume in response to haemorrhage or in pulmonary congestion. The figure of 7.78% of total cardiac output obtained for hepatic arterial flow in the liver by Brookes (1970) is comparable to the value that can be calculated for hepatic artery flow from the results of Greenway and Stark. Electron microscopy of the terminal branches of the hepatic artery indicates that there are well-developed smooth muscle cuffs that act as pre-capillary sphincters. The muscle cells of these sphincters are innervated by non-myelinated autonomic nerves (Burkel, 1970).

The interrelationship between the portal venous and the hepatic arterial blood supply of the liver has been studied by Bloch (1955) and McCuskey (1966) by in-vivo techniques in a variety of animals. The main point of difference in the two sets of findings is regarding the existence of arterio-portal anastomoses. Bloch described them as occurring between the two sets of vessels via short connections arranged like the rungs of a ladder whilst McCuskey found no direct anastomoses of this kind but described the hepatic arterial vessels as entering directly into the hepatic sinusoids near their origins from the portal vein radicles. He considered that functional arterio-portal anastomosis could occur in the presence of back-flow from the hepatic sinus into the portal vein radicle. Both investigators found that the control of hepatic sinus blood flow was achieved by the interaction of inlet and outlet sphincters present at each end of the sinusoids; they are also found at each end of the intersinusoidal sinusoids identified by McCuskey. The pressure drop that occurs as the blood flows through the sinusoids that course between the leaflets of hepatic parenchymal cells to enter the hepatic vein radicles situated at the centre of each hepatic lobule has been measured in the frog's liver as 3.8 mmHg.

The mingling of the portal venous and the hepatic arterial blood achieves two desirable results. One is that the oxygen-poor portal blood bringing the absorbed products of digestion from the alimentary canal to the liver has a proportion of oxygen-rich blood mingled with it before it traverses the hepatic lobules which rely solely on the sinusoidal blood flow for their respiration. The other is that the portal blood which is at a venous level of pressure will be energized by the inflow of the high pressure arterial blood in the manner of the jet pump principle used in hydraulics. To what extent the well-documented rhythmic contractile activity of the portal vein–anterior mesenteric vein system (see p. 19) may contribute to the propulsion of blood within the portal circulation of the liver is not clear. Recently, Tsáo, Glagov and Kelsey (1970), as a result of an electron microscopic study of the portal vein, postulated a propulsive role for this vessel, and Helfer and Jaques (1970) have measured the variations in venous output of mouse ligated portal vein sac preparations by a micromanometric technique. To what degree the valves that are present in the intra-abdominal veins and their tendency to regress with age (Kampmeier and Birch, 1927) might influence such a propulsive action is not clear.

The presence of a portal vascular system has been confirmed in the pituitary of mammals (Akmayev, 1971) and of birds (Mikami et al., 1970) In these cases the vessels of the median eminence of the posterior pituitary supply a portal circulation to the anterior lobe of the pituitary. Worthington (1962) has given a most detailed and interesting account of this highly important circulatory arrangement. Ferrer (1957) has correlated the distribution of pituitary basophile cells with particular vascular patterns within the adenohypophysis.

A renal portal system is present in all vertebrates with the exception of the cyclostomes and the mammals (Romer, 1950). The function of this system is to distribute the venous blood returning from the posterior parts of the body to a greater or lesser extent through the peritubular capillaries of the kidney. Akester (1967) has studied the varying distribution of radio-opaque materials injected into the external iliac veins of the domestic fowl and their entry into the renal portal system which is controlled by the action of richly-innervated muscular renal portal valves (Akester and Mann, 1969) which they consider unique vertebrate intravascular structures. Bennett and Malmfors (1970) have demonstrated very rich adrenergic innervation of the vessels of the renal portal system in the fowl. By selective action of the valving mechanisms in this system

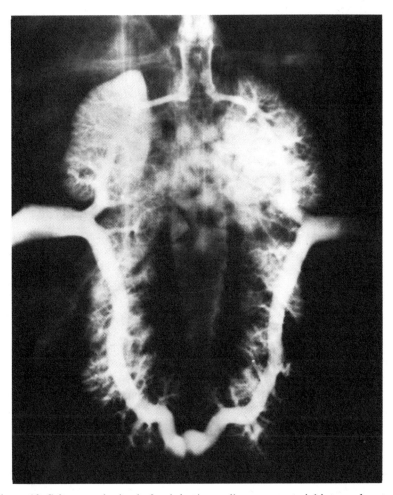

Figure 10. Sciagram obtained after injecting radiopaque material into each external iliac vein of a domestic fowl. These veins can be seen entering the figure at the right and left margins. They bifurcate into the cranial (upper) and caudal (lower) renal portal veins. The outlines of the kidneys can be clearly seen due to the radiopaque material being distributed throughout the peritubular capillaries. Connection with the vertebral venous sinuses occurs near the top of the figure. (Akester, 1967.)

venous blood can be routed to enter the kidneys or to by-pass them and drain either directly into the inferior vena cava or into the hepatic portal system or into the vertebral venous sinuses (figure 10).

Physiological evidence for the activity of the muscular pulmonary vein walls can be found in a study of the increased pulmonary vascular resistance occurring during hypoxia in dogs, from which Furnival, Linden and Snow (1970) concluded that the pulmonary veins contributed 18% of the increase in resistance. It can be inferred also from the work of

Aarseth (1970) who found that pulmonary blood volume was selectively and considerably reduced following haemorrhage in rats. The pulmonary veins have been demonstrated to have only adrenergic innervation (Ehinger, Falck and Sporrong, 1966; Fillenz, 1970; Bennett, 1971) in contrast to the pulmonary artery branches which have both adrenergic and cholinergic innervation (Silva and Ikeda, 1971). Bennett found that the innervation of the pulmonary veins was considerably denser in proximal than in distal regions.

SINUSOIDAL VESSELS

Attention has so far been drawn to the essentially closed nature of the vertebrate circulation as opposed to the open system found in most invertebrates. There are however certain organs in vertebrates that are supplied with a sinusoidal form of circulation. Sinusoids differ from capillary exchange vessels in that they are wider, more irregular in size and shape and have discontinuous or even no endothelial linings. In the Myxini, such sinusoidal circulatory patterns are widespread throughout their tissues. The sluggish-flow low-pressure configuration of this type of system is reflected in the presence of a series of venous hearts which act to help return the blood from the tissues into the venous system.

Examples of tissues in which sinusoidal vascular patterns are generally quoted as persisting in mammals are liver (Beard and Beard, 1927; Bloch, 1955; Majno, 1965; McCuskey, 1966, 1968; Wisse, 1970; Grubb and Jones, 1971; Widmann, Cotran and Fahimi, 1972), spleen (Beard and Beard, 1927; Williams, 1950; Majno, 1965; Snodgrass and Snook, 1971), bone marrow (Beard and Beard, 1927; Majno, 1965; De Bruyn, Breen and Thomas, 1970) and adrenal gland (Beard and Beard, 1927). All these tissues, together with the lungs (Wislocki, 1924), have the property of avidly removing colloidal materials injected into the blood stream (Wislocki, 1924; Beard and Beard, 1927; Cappell, 1929a, b; Williams, 1950; Bloch, 1955; Kelly, Brown and Dobson, 1962; Snodgrass and Snook, 1971; Widmann et al., 1972). This ability to take up materials from the blood stream is a feature of the reticulo-endothelial system which also includes, in addition to the sites already mentioned, the cells lining the lymph sinuses of the lymph nodes. The tendency for materials to be removed from the blood stream in organs with sinusoidal blood vessels has been linked with the greater incidence of metastatic tumor growths that occur in these organs (Boyd, 1953; Cappell, 1958). The converse, namely the entry of cells into the circulation from regions of haematopoiesis, as studied by McCuskey (1968) in the foetal liver, also occurs into sinusoidal vessels. Red blood cells were seen to penetrate the sinusoidal endothelium and in the process they left behind their nuclei which were promptly phagocytosed by the related reticulo-endothelial cells. In general

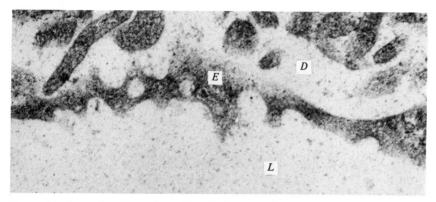

Figure 11. Part of rat liver sinusoid with fenestrated continuous endothelial lining (*E*) with 'sieve-plate' character separating the lumen (*L*) from the Space of Disse (*D*). Magnification ×65000.

the sinusoidal vessels have been considered to possess discontinuous endothelial linings characterized by intercellular gaps (Bennett, Luft and Hampton, 1959; Majno, 1965).

Recent electron microscopic studies of the liver have demonstrated quite clearly, however, that the sinusoids are lined by a fenestrated continuous endothelium (Wisse, 1970; Grubb and Jones, 1971; Widmann *et al.*, 1972), picturesquely described as 'sieve-plates' (figure 11). This form of endothelium could well supply the high permeability characteristics of the hepatic circulation (Mayerson *et al.*, 1960; Greenway and Stark, 1971) which allows passage of molecules up to 412000 mol. wt., corresponding to a molecular diameter of ~ 14 nm. The presence of fenestrated sieve-plates rather than simply endothelial gaps or discontinuities may serve to provide some biological control over the high permeability of the vessels and help to resolve the paradox, pointed out by Greenway and Stark (1971), of having extremely permeable vessels but a low albumin content in the extracellular fluid. Wisse (1970) has raised the possibility that the fenestrae of the sieve-plates lining the hepatic sinusoids may exercise some selective effect on the uptake of chylomicrons of varying sizes from the blood by the liver.

Electron microscopy has also clarified the phagocytic function of the lining cells of the hepatic sinusoids. The highly phagocytic Kuppfer cells of the sinusoids are a completely distinct population of cells that can be readily distinguished from the true endothelial cells (Wisse, 1970; Widmann *et al.*, 1972). The Kuppfer cells make up 40% and the endothelial cells 48% of the population of cells lining the sinusoids. The Kuppfer cells are highly phagocytic and have high levels of endogenous peroxidase activity, whilst the endothelial cells are weakly phagocytic and are uniformly lacking in endogenous peroxidase activity. The Kuppfer cells have been shown to be an actively dividing population of cells by [^{3}H]thymidine labelling experiments (Kelly *et al.*, 1962). A high turnover of Kuppfer cells is supported by the findings of Smith, McIntosh and Morris (1970) who detected these cells in large numbers in the hepatic lymph of sheep following intravenous injection of colloidal gold or carbon. Also highlighting the dynamic nature of the Kuppfer cell population is the evidence obtained by Howard, Boak and Christie (1966) that approximately two-thirds of the dividing liver macrophages following bacterial stimulation were derived from thoracic duct cells in radiation chimaeras.

Despite the statements made by various investigators of the structure of the splenic sinusoids (p. 26) the actual arrangement of the circulation within the spleen is extremely complex. The calibre of the splenic artery is out of all proportion to the size of the organ, being almost as wide as the femoral artery (Brash and Jamieson, 1947). Its final ramifications within the spleen are in the form of the highly characteristic penicillate arteries that have nodular sphincter-like collections of smooth muscle cells in their walls. These specialized vascular muscle collections act as stopcocks to control the blood flow within the spleen. These vessels lie at the centres of the aggregates of lymphoid cells known as Malpighian bodies. These features are readily recognized by all investigators and were seen *in vivo* by Williams (1950) who studied splenic autografts in rabbit ear chambers (Sandison, 1924; Clark *et al.*, 1930). Splenic autografts had a highly individual form of circulation and were the only autografts found to increase their size after establishment in ear chambers, in fact by a factor of three (Williams, 1951). He identified splenic sinusoids which were extremely labile structures continuously forming and disappearing and were described in regions as being 'multilocular labyrinthine channels'. In general it appeared that the sinusoids did have an endothelial lining which was unusual in being very thin and having large nuclei projecting into the lumen. However, sinusoids with intact identifiable walls could quite definitely communicate with others with no identifiable walls. Williams suggested that the spleen should be thought of as a large chamber interposed in the blood stream which has certain specialized functions and is in fact a specialized part of the vascular system, in which case the question of whether or not the sinusoids have an endothelial lining becomes irrelevant (Williams, 1950). In-vivo studies on the circulation in the mouse's spleen were made by Parpart, Whipple and Chang (1955) who found an amazingly active and ever-changing intermediate circulation. They found that the vast majority of arterioles terminated by spewing their blood through funnel-shaped openings into large pulp spaces with no detectable walls. The blood drained from these pulp spaces into veins by way of slits in the vein walls. A small number of arterio-venous anastomoses existed and a small number of arterioles fed loose irregular capillary networks which drained into venules and veins in the usual manner. M. H. Knisely in the discussion appended to this paper takes issue with every point and considers that the splenic sinusoids are anatomical entities with definite cellular linings. Snodgrass and Snook (1971) discussed very fully the questions relating to the nature of the intermediate circulation in the spleen and from their work concluded that a unique and intimate relationship existed between the highly phagocytic cells of the reticulo-endothelial system and the terminations of the arterial capillaries in the spleen. The sinus-lining cells had a fusiform shape and they considered that they were a functionally different population from the reticulo-endothelial cells. Opdyke (1970) studied changes in the outflow rates and haematocrits from exteriorized dog spleens in response to sympatheticomimetic drugs. He concluded that the results of his experiments were compatible with the view that blood flow in the spleen occurred through channels between tightly packed masses of red cells that did not possess endothelial linings. The role of the spleen in providing red cell storage relies on its ability to contract and on the ready access of the stored cells into the splenic vein radicles. For a further discussion of the nature of the splenic intermediate circulation see Selkurt (1962).

The general question now arises of blood flowing in direct contact with tissues, as propounded above in the spleen. There is no doubt that this occurs in the haemochorial type of placenta found in the primates (Ramsey, 1962). In this situation the chorionic villi of foetal origin penetrate the maternal blood vessels that come to form lakes of blood bathing the villi and the foetal trophoblast comes to serve the function of endothelium (Ramsey, 1955). This surely is one of the most remarkable examples of exemption from the normal course of immunological reactions. Whilst the haemochorial placenta has reduced the number of layers separating the maternal and foetal blood streams to only three (trophoblast, mesoderm and foetal endothelium) it is not necessarily the most efficient form of placenta for oxygen exchange. Lovatt Evans (1949) points out that the counter-current arrangement of maternal and foetal capillaries, as found for instance in the rabbit placenta, is the most efficient type for such exchange to occur.

Finally, there is the possibility that a sinusoidal form of circulation may persist in the mammalian heart. This proposition has been brought forward recently by Hammond and Moggio (1971) who described a system of nutrient sinusoidal vessels that meander through the myocardium of the left ventricle carrying blood from the coronary arteries to empty directly into the chamber of the left ventricle. They raise the question whether the sinusoidal vessels of the mammalian heart are merely the remnants of the early phylogenetic development of the heart (Stehbens and Meyer, 1965), or whether in fact they represent a form of circulation with an important physiological role in the myocardium. Wearn *et al.*

30 The vessel lumen

(1933) give a good account going back to 1705 of the history of these putative structures in the heart. I feel that the descriptions and proposed roles of these myocardial sinusoidal vessels in mammals are not adequately defined at the moment, but certainly merit detailed investigation in the future.

3

THE ENDOTHELIUM

The word 'endothelium' is a relatively modern term being derived from the Greek *endo* = 'within' or 'lining' and *thēlē* = 'nipple'. How this word has arrived at its accepted modern meaning, as restricted to the smooth cellular membrane forming the inner lining of blood and lymphatic vessels, I do not know. Endothelium is a generic term just as the term epithelium is. It is not possible to talk of 'the endothelial cell' as some single entity, for endothelium can differ from vessel to vessel and from organ to organ. In fact Majno (1965) has made the point that there are as many types of capillary as there are tissues. This of course is not solely in terms of their endothelium, including as it does structural, histochemical and haemodynamic features of each capillary, but it does serve to stress the concept that any organ, functioning within the body, cannot be considered as being solely a parenchyma which merely happens to need a blood supply. In any organ there has to be a perfect meshing of its two major components, the parenchyma and the vasculature.

Descriptions of living endothelial cells have been made from in-vivo observations (Sandison, 1932; Clark and Clark, 1932, 1935, 1943; Williams, 1954) and from observations in tissue culture (Maximov, 1916, 1917; Willmer, 1945; Woodard and Pomerat, 1953; Farnes and Barker, 1963). General descriptions of endothelial fine structure have been made by Bennett, Luft and Hampton (1959), Fawcett (1959, 1963), French (1963), Majno (1965) and Florey (1966).

CELL JUNCTIONS

The general shape and interrelations of endothelial cells are best appreciated from light microscopic techniques that use the method of silver deposition and development (Poole, Sanders and Florey, 1958) to delineate the margins of each cell within an endothelial membrane (figure 12). In the past some doubt existed as to whether the deposition of silver actually did correspond to the cell margins. This criticism can now be largely discounted in view of the findings of electron microscopy (Buck, 1958; Florey, Poole and Meek, 1959). It would seem reasonable to suppose that silver ions, each bearing a single positive valence, are being chemically

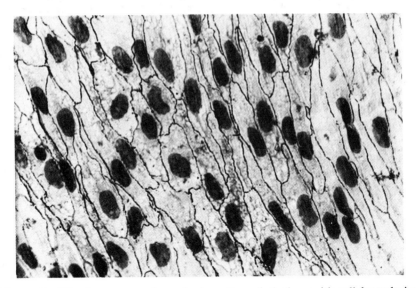

Figure 12. Häutchen preparation of rat aortic endothelium with cell boundaries stained with silver. Magnification × 400 (approx.). Micrograph supplied through the kindness of Dr J. C. F. Poole.

bound to negative ions, presumably halide, localized in these regions, the subsequent formation of metallic silver deposit being due to the photochemical properties of silver halide compounds. In this regard it is interesting to note that negatively charged materials such as heparin and various sulphonic acids are concentrated at the endothelial junctional regions where they can be demonstrated by the appropriate staining reaction (Florey *et al.*, 1959; Ohta *et al.*, 1962). The reason why there should be preferential localization of substances at these junctional regions is not clear. Numerous publications have sought to prove the existence of a cement substance that is responsible both for joining together the cells of the endothelial membrane and for determining its permeability characteristics (see Chambers and Zweifach, 1947). Both Buck (1958) and Florey *et al.* (1959) in examining the question of silver binding found no evidence for a 'cement' substance. The absence of cement substance has been stressed also by Fawcett (1963). There is no theoretical necessity for a cementing substance for like cells to form strong adhesions one with another (Curtis, 1960, 1964). As pointed out by Florey *et al.* (1959) the more generalized ability of the endothelial cell surface to bind materials is basically more interesting than the old idea of localized regions of cement substance. This binding to the endothelial luminal surface may be due to the endo-endothelial coat demonstrated electron microscopically by Luft (1965) and by Behnke and Zelander (1970) who inferred that the binding of positively-charged Alcian blue in their experiments indicated the

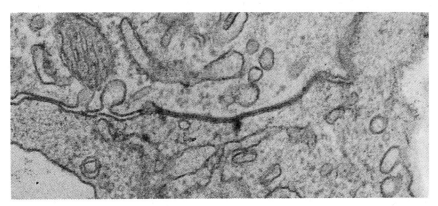

Figure 13. An oblique section through an endothelial cell junction in the rat's aorta. One large and one small region of zonula occludens type of intercellular junction can be identified. Magnification × 80000. Electron micrograph supplied through the kindness of Dr R. G. Gerrity.

presence of negatively-charged substances in an endo-endothelial layer. Easty and Mercer (1962) postulated that cell surfaces in general may have a negatively-charged mucopolysaccharide coating which would account for many of their observed properties, a view also propounded by Bennett (1963) in the glycocalyx concept of the cell surface. Berlepsch (1970) found evidence of mucopolysaccharide production by endothelial cells as also did Curran (1957) using ^{35}S incorporation studies in endothelium. However, Stehbens (1962) in repeating the ^{35}S studies could find no evidence of mucopolysaccharide synthesis by endothelial cells. The possibility that the endo-endothelial coat may be a layer of adsorbed fibrin with a cementing role has also been suggested (Copley and Scheinthal, 1970).

The endothelial cell junctions of various blood vessels have been studied by electron microscopy (Muir and Peters, 1962; Majno, 1965; Karnovsky, 1967; Bruns and Palade, 1968a). There is now general agreement on the existence of a region of tight junction or membrane fusion occurring in practically all junctions between endothelial cells in which the two triple-layered unit membranes (Robertson, 1959) of the related cells merge their outer leaflets to produce a quintuple-layered membrane (Muir and Peters, 1962) (figure 13). This region of membrane fusion is always towards the luminal end of the junctional region and is termed either a *zonula occludens* (Majno, 1965; Karnovsky, 1967; Bruns and Palade, 1968a; Giacomelli, Wiener and Spiro, 1970) if it is considered to form a band of attachment completely around the cell, or else a *macula occludens* (Karnovsky, 1967; Giacomelli *et al.*, 1970) if such regions are considered to be discontinuous 'buttons' of attachment disposed at intervals around the margins of the cells. In fact Karnovsky (1967) considered that zonulae occludentes are only present in cerebral vascular

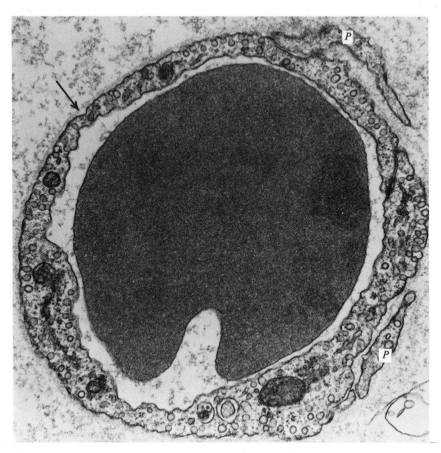

Figure 14. A very narrow nutrient vessel in the rabbit's myocardium containing a red cell whose profile suggests the bullet or parachute cell shapes observed *in vivo* (cf. figure 4). The vessel is composed of a single endothelial cell which forms a junction with itself near the mid right margin of the figure. Small profiles of pericytes are present (*P*). A basement membrane (arrow) is narrowly separated from the endothelial surface. Magnification × 30000.

endothelial junctions and that all other endothelia have discrete maculae occludentes at their junctions. This modern Latin terminology is derived from a paper by Farquhar and Palade (1963) in which they very elegantly characterized the basic components of epithelial cell junctions. (See also p. 54 for evidence of a gap being present in this type of junction.)

The endothelial attachment regions related abluminally to these tight junctions are characterized by strict parallelism of the two related plasma membranes which are separated by distances of ∼ 20 nm. This appearance conforms to the *zonula adhaerens* component of intercellular junctions described by Farquhar and Palade. True desmosomes are complex

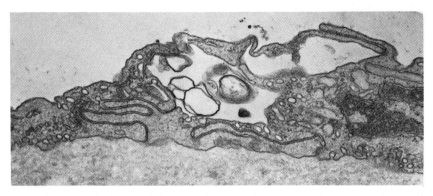

Figure 15. An endothelial cell junction in the aorta of a rat. There is complex interdigitation of the opposing cell membranes and, in addition, the luminal endothelial surfaces have formed a series of flaps related to the cell junction which enclose a quantity of vesicular material. Magnification ×27500.

structures normally found in epithelial junctions and do not occur in endothelial junctions of the higher vertebrates (Bruns and Palade, 1968a). According to Fawcett (1963) they are present, however, in the endothelial intercellular junctions of the blood vessels of the rete mirabile of the teleost swim bladder. The use of the term 'desmosome' in connection with mammalian endothelial junctions is however still sometimes used to this date and can only be deplored. The relationship of vascular permeability to the structure of endothelial junctions will be considered below.

The relationship of endothelial cells to one another at their junctions may show considerable variety. In the simplest cases there is a direct abutting of the margins of the related cells over relatively short and straight courses (figure 14). From this very simple arrangement progressively more complicated forms of interrelation of the cell margins may be seen (figure 15). Rhodin (1967) found that in the endothelial junctions of arterioles the margin of the upstream cell generally overlaps on the luminal side the margin of the downstream endothelial cell. Such an arrangement ensures that the blood stream in the vessel assists in maintaining the seal formed between the two endothelial cells. Mention must be made here of 'seamless' blood vessels observed quite frequently by Schoefl (personal communication) (figure 16) and by Wolff (1964) which are composed of a single endothelial cell with no apparent cell junction present. These vessels are quite distinct from those cases, more frequently encountered, where an endothelial cell is rolled upon itself to form a single true endothelial junction (Kisch, 1957; Fawcett, 1959) (figure 14).

By using the technique of '*en face*' sectioning of an endothelial membrane the very complex interdigitations of the related endothelial cell margins can be appreciated (Gerrity and Cliff, 1972) (figure 17). Another

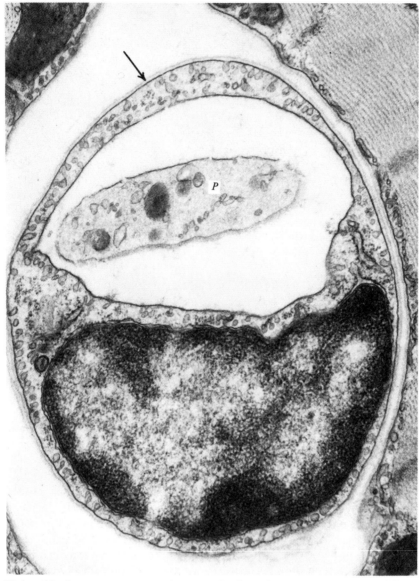

Figure 16. A very narrow nutrient vessel of murine myocardium containing a single blood platelet (*P*) in its lumen. The vessel wall is composed of a single endothelial cell with no junction. A basement membrane (arrow) is narrowly separated from the basal surface of the endothelium. Magnification ×30000. Electron micrograph supplied through the kindness of Dr G. I. Schoefl.

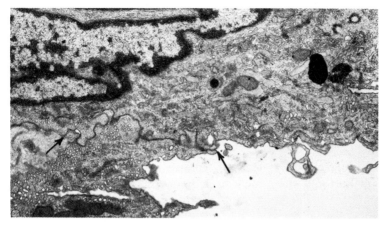

Figure 17. Rat aortic endothelium sectioned *en face*. An endothelial cell junction follows a serpentine course between the two arrows. Note the paired centrioles and associated Golgi membranes at the top right corner of the figure. Mitochondria and Weibel and Palade bodies are numerous. Magnification × 16000.

way of demonstrating these complex interrelations at high resolution has been by the application of freeze–fracture or freeze–etching techniques (Staehelin, 1968) to the study of blood vessels (Leak, 1971).

CYTOPLASMIC PROCESSES

All techniques used to study the endothelial membrane reveal the existence of fine endothelial cytoplasmic processes that extend into the vessel lumen, particularly in relation to the junctional regions (figure 15). In certain instances these projections are in the form of fine finger-like microvilli extending from the endothelial surface into the lumen (Kisch, 1957; Morato and Ferreira, 1957; Policard, Collet and Pregermain, 1957; Policard and Collet, 1958; Fawcett, 1959, 1963; Williamson and Grisham, 1961). Certain other projections have the form of flaps or folds extending from the endothelial surface (Fawcett, 1963; Ludatscher and Stehbens, 1968; Mikami *et al.*, 1970; Gerrity and Cliff, 1972). Leak (1971) has confirmed the presence of intraluminal cytoplasmic protrusions, particularly in relation to endothelial junctions, using the freeze–etch technique. The function of these intraluminal projections and the reason for their greater localization on the endothelial surface related to cell junctions is obscure. Possible functions suggested for the microvilli are absorptive (Policard and Collet, 1958) and the trapping of leukocytes during inflammation (Williamson and Grisham, 1961). The surface folds or flaps on the luminal surface of endothelial cells are more constantly related to cell junctions. Fawcett (1963) has suggested that their presence in vessels which are extremely permeable to water, as in the choroid rete of fish, or in induced cerebral oedema in rabbits, may indicate that they have a true pinocytotic function in taking up and passing fluid through the endothelial membrane from the plasma to the tissues. Mikami *et al.* (1970) considered that the highly developed endothelial flaps present in the vessels of the hypophysial portal system probably have valve-like actions in these vessels. The great increase in the aortic endothelial surface area produced by the flaps and folds that become more numerous with age (Gerrity and Cliff, 1972) are considered to be related to increased permeability of this membrane.

Highly developed systems of fronded cytoplasmic extensions of the endothelium into the lumen have been observed by Irey, Manion and Taylor (1970) in the blood vessels of women dying with various thrombotic lesions whilst taking oral contraceptives. Irey *et al.* considered that the endothelial changes were very probably related to the steroids being taken before death. There is increased platelet consumption associated with the intravascular proliferation of endothelial cells that break the blood stream up into numerous additional channels in the condition known as diffuse angio-endotheliosis (Meadors and Johnson, 1970). Whilst this may be a purely mechanical effect, rather analogous to the increased platelet consumption that is found in association with prosthetic cardiac valves (Harker and Slichter, 1970), it more probably reflects some basic interrelationship that exists between endothelial cells and blood platelets.

The relationship between endothelial cells and platelets in certain diseases characterized by increased capillary fragility has been considered by various authors (Quick, 1957; Johnson *et al.*, 1964; Nour-Eldin, 1966). Johnson *et al.* (1964) found evidence for a trophic relationship in that platelets were apparently taken up by and incorporated into endothelial cytoplasm. Marchesi (1964) considered that a trophic relation existed in the course of the inflammatory reaction whereby the development of the acid phosphatase activity observed in endothelial cells was derived from blood platelets incorporated into their cytoplasm. However as Marchesi considered in arriving at his conclusions that endothelial cells did not normally possess acid phosphatase activity, the recent demonstration by Ts'ao (1970) of acid phosphatase in normal arterial and venous endothelium must make these conclusions doubtful. Gore, Takada and Austin (1970) produced severe platelet reduction in guinea pigs by means of an antiplatelet antibody injected intraperitoneally. They found that the myocardial vessels had large intercellular gaps and that this change could be rapidly reversed by administration of fresh blood platelets. They concluded that the supporting effect of platelets on the endothelium was through the maintenance of tight intercellular junctions. Paradoxically Margaretten and McKay (1971), in producing

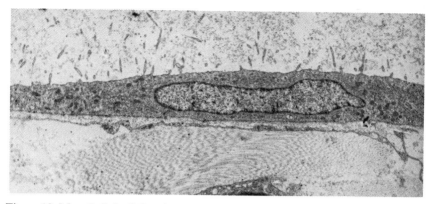

Figure 18. Mesothelial cell forming serosal surface of rabbit pericardium. Magnification × 5000.

severe depression of platelet counts in rabbits by intravenous injection of an antiplatelet antibody, found that blood platelets were apparently essential for the development of the Arthus reaction in supplying a permeability factor that would allow the exit of circulating antibody to react with the challenging antigen (bovine serum albumin in this case). The extreme importance of preparing an antibody specific for platelets which does not react with any shared antigens on endothelial cells in such experiments is stressed by the work of Adelson, Heitzman and Fennessy (1954) who reported a condition termed 'thrombo-hemolytic thrombocytopenic purpura'. They concluded that this represented a hypersensitivity state that involved red blood cells, platelets, megakaryocytes and blood vessel walls. Finally, the difficulty encountered in obtaining adequate tissue culture preparations of endothelial cells may well involve the absence of important trophic effects postulated for the blood platelets.

CELL SURFACES AND CELL SHAPE

The membranous sheet-like character of the endothelium makes it immediately obvious that each cell is in a highly improbable form, in physical parlance, with an extremely high surface-to-volume ratio. This characteristic of endothelial cells is also shared by the mesothelial cells that line the pleura, the pericardium, the peritoneum and joint synovia (Willmer, 1945) (figure 18). In these situations the usual mesodermal form of cellular distribution and interrelation, as exemplified in the connective tissues (Jacobson, 1953), has been modified to produce epithelial-type structures.

There are a number of agencies that could possibly be operating, either alone or in combination, to maintain the high surface energy state of each endothelial cell. From this point of view it must be realized that for thermodynamic reasons an interfacial system will always tend to assume a lower-surface-energy form (Glasser, 1944; Manly, 1970). This is well illustrated in the 'rounding-up' of cells that occurs as a sign of impaired vitality or of cell death. The in-vitro behaviour of macrophages, which tend to be rounded or spherical in form until supplied with energy in the form of ATP or ADP which is utilized by their cell surfaces (North, 1966a), is the same phenomenon acting in the reverse direction. Such cells flatten out and spread dramatically on the surface of the slide to attain high-surface-energy configurations very reminiscent of endothelium so long as an exogenous supply of energy, as ATP, is maintained. Once this is exhausted the macrophages rapidly return to a rounded state (North, 1967). Here, then, is one possible agency that could be operating to maintain the high surface energy of the endothelium, namely the utilization of chemical energy by the endothelial cell surfaces. Endothelial cells have been demonstrated to possess ATPase activity, mainly in their surface caveolae (Marchesi and Barrnett, 1963; Hoff and Graf, 1966), which is an interesting feature of similarity with macrophages which possess strong ATPase activity at their cell surfaces. As a corollary of this, any impairment in the energy-producing metabolic functions of the endothelium would tend toward the rounding-up or 'swelling' of the endothelial cells. In fact 'swelling' of endothelium is frequently observed occurring very rapidly in response to various injuries or the action of noxious agents including those associated with autografting (Lund and Jensen, 1970a), pharmacological stimulation at high dose levels (Constantinides and Robinson, 1969a), anti-endothelial antibodies (Clark and Jacobs, 1950), Rickettsial infection as in typhus (Manion, 1963) and virus infection as in equine viral arteritis (Estes and Cheville, 1970).

In addition to this energy-consuming process, other processes that may act to help maintain the extremely flattened shape of the endothelial cell must be considered. One such is the agency of surface-active materials which would operate to lower the surface tension at the cell's luminal and abluminal surfaces. Surface-active agents have detergent-like properties and concentrate spontaneously at interfacial regions (Glasser, 1944). Both proteins and phospholipids are examples of naturally occurring surfactants. Probably the most important biological example of surfactant material is that which occurs in the lung. The contribution of interfacial tensile forces to the total wall tension of blood vessels has been considered by Burton (1954). Whilst there is no objective proof of the presence of a layer of surface-active material on endothelial surfaces there are certain suggestive findings which would support the proposition. A definite layer of adsorbed material which appears to be, at least in part, of a negatively-charged polysaccharide nature has been demonstrated on the endothelial luminal surface (see p. 32), which would have no surface-active proper-

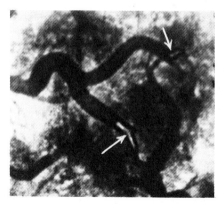

Figure 19. In-vivo photomicrograph showing two air bubbles (arrows) lodged within branches of an arteriole in a rabbit ear chamber. The meniscus formed at the blood–air interface demonstrates the wettable character of the lining endothelium. Magnification × 20.

ties. However, Whiffen and Gott (1964) demonstrated that cationic surface-active agents bound heparin (a negatively-charged polysaccharide) to hydrophobic graphite–water interfaces, and suggested that a similar situation might obtain with endothelial surfaces.

Attempts to determine the actual surface properties of endothelium are not numerous in the literature. Glasser (1950), in describing the gas-bubble contact-angle technique for characterizing the nature of the aqueous boundary at various types of surfaces, stated that the endothelial surface observed in the rat's mesentery was hydrophilic. My own recent observations on rabbit vessels are in agreement with this (figure 19). By way of direct contrast Nichol et al. (1951) demonstrated that the endothelial surface was normally hydrophobic. The possibility of the endothelium altering its surface wettability characteristics has been proposed as a factor in haemostatic events (Quick, 1957; McKay and Hardaway, 1963). It would appear that further in-vivo observations aimed at clarifying this important property of endothelium and its possible involvement in the vascular changes associated with haemostasis, thrombosis and acute inflammation would be of value.

In all considerations of the endothelial cell it is well to remember that it has two different surfaces, the luminal and the abluminal. Apart from the obvious anatomical differences in their relation to the lumen, basement membrane system and so on, there are other profound differences related to their configuration. The luminal plasma membrane surface is concave with respect to the exterior of the cell whilst the abluminal plasma membrane surface is convex with respect to the exterior of the cell. These different surfaces of essentially lipid membranes (Robertson, 1959; Stoeckenius, Schulman and Prince, 1960) have, respectively, the con-

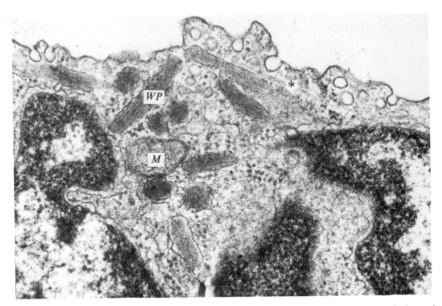

Figure 20. An endothelial cell of the rat's aorta contains numerous rod-shaped Weibel and Palade bodies (*WP*) cut both obliquely and in cross-section. Microtubular structures are present both within these bodies and free in the cytoplasm (∗). A single mitochondrion is present (*M*). Magnification ×50000.

figurations of water-in-lipid and a lipid-in-water emulsion interface. The separation of the hydrophilic polar groups present at the convex membrane surface of the abluminal side will result in a lower net surface charge density which will tend to increase intercellular adhesion (Curtis, 1960). The crowding together of polar groups on the concave luminal surface will have the opposite effect by increasing the net surface charge density. Such differences in membrane external surface properties of the endothelial cells in fact impose a radial polarization on the endothelium as a whole. Such polarization may account for the one-way traffic that occurs in cell migration across the endothelial membrane. Leukocytes leave blood vessels very readily in certain sites and situations but they do not re-enter them again from the tissues.

Other agencies which may be operating to help maintain the flattened endothelial shape are in the nature of physical 'splinting'. The most obvious candidate for such a function is the basement membrane which surrounds the endothelium and which forms contacts with the more robust collagenous and elastic elements either of the vessel wall or, in the case of the exchange vessels, of the surrounding fibrous tissue stroma. The adventitial cells, or pericytes, of the smaller vessels which correspond, in general, to the tunica media of larger vessels (Zweifach, 1937; Clark and Clark, 1940; Krogh, 1959) are invested by basement membrane systems which merge with those of the endothelial cells and thus can also contribute to some form of splinting support for the endothelium. In the case of certain larger vessels definite fibrillar interconnections with apparent anchoring or supportive functions can be identified running between the endothelial cells and the underlying structures of the blood vessel wall (Ts'ao and Glagov, 1970a; Gerrity and Cliff, 1972). There may also be operating various 'internal' splinting mechanisms (Holfreter, 1947) within the cytoplasm of the endothelial cells. These cells possess both microfibrils and microtubules (figures 20, 30). Microtubules, in particular, are considered to have cytoskeletal or supportive functions, as demonstrated in macrophages (Allison, Davies and De Petris, 1971; Bhisey and Freed, 1971). The observations on the in-vivo properties of endothelium by Chambers and Zweifach (1944) who used a microneedle to raise a spike from the endothelial surface indicate that endothelial cytoplasm has sufficient intrinsic viscosity to cause the spike to persist for several seconds after it is formed. Whilst the tracts of microfibrils that are present in the endothelium will cause increased viscosity of the cytoplasm, their other possible role as contractile elements will be discussed in conjunction with the microcirculation (see p. 154).

ORGANELLES AND INCLUSIONS

Certain peculiar inclusions are found in vertebrate endothelial cells and termed 'Weibel and Palade bodies' (Weibel and Palade, 1964; Burri and Weibel, 1968; Gimbrone, Cotran and Folkman, 1974). These structures are basically packets of microtubular material enclosed within unit membranes. They arise in relation to the Golgi apparatus (Sengel and Stoebner, 1970) and occur more frequently on the luminal rather than the abluminal side of the cell. They are elongated, often curved structures which in some instances are twisted together in the cytoplasm like a mass of worms (figures 17, 20). These bodies measure 0.1 μm in diameter by 3.0 μm long. The material of moderate electron-density which they contain represents some form of secretion elaborated by the endothelial cell. Burri and Weibel (1968) pointed out the similarity between these endothelial inclusions and the α-granules of the blood platelets which have thromboplastic activity. They also showed that incubation of rat aortic strips with epinephrine (adrenaline) (5×10^{-9} gml^{-1}) produced a 40% reduction in the number of these bodies as compared to Ringer-incubated control strips. They suggested that the bodies were expelled towards the vessel lumen. However the nature of the secretion from endothelial cells varies from species to species and even from vessel to vessel within the same animal. Astrup and Buluk (1963) emphasized the greatly different methods employed in various species of mammals to control mural thrombosis, a process considered by Duguid (1954) to be of crucial aetiological importance in the development of atherosclerosis in human beings. Astrup and Buluk showed that the control of this phenomenon in the dog was through lack of tissue thromboplastin, in the rat by the presence of an anticoagulant, in the monkey by the presence of fibrinolytic activity and in man through the agency of a powerful latent fibrinolysin. Coccheri and Astrup (1961) showed that in human beings the arteries had higher thromboplastic and lower plasminogen fibrinolytic activator systems than the veins. The possibility that two such systems, the one producing fibrin intravascularly and the other removing it, are working in balance in the circulation is suggested by the results of McKay and Hardaway (1963) who showed that the half-life of fibrinogen in the circulation was of the order of 6 days. Warren (1963) has demonstrated fibrinolytic activity of endothelium from a variety of tissues using the ingenious fibrinolytic autograph technique originally devised by Todd (1959). Ashford, Freiman and Weinstein (1968) showed by using ϵ-amino caproic acid, which antagonizes plasminogen activator, that the higher levels of plasminogen present in small veins as compared with large ones were most probably due to the higher surface-to-volume ratio of the former and not to increased intrinsic activity of their endothelium.

The Golgi apparatus has been mentioned in association with the Weibel and Palade bodies (above) and it is claimed by Sengel and Stoebner (1970) that these bodies arise in the Golgi system of membranes. The involvement of the Golgi apparatus in the packaging and concentration of secretion products is well supported in the literature (Oberling, 1959; Caro, 1961; Munger, 1961; Zeigel and Dalton, 1962; Wissig, 1963; Caro and Palade, 1964). In general, the Golgi apparatus is considered to be poorly developed within endothelial cells (Fawcett, 1959; Rhodin, 1962a; French, 1963). However, in immature endothelium, as found in neonatal animals, in repair or granulation tissues and also in aging endothelium of the rat's aorta, this is not the case and this apparatus may occupy a large part of the cytoplasm related to the nucleus of endothelial cells (Cliff, 1963; Gerrity and Cliff, 1972) (figures 21, 22). The Golgi apparatus is composed of stacks of paired parallel smooth-surfaced membranes associated with membrane-bounded vesicles and free ribosomes in the cytoplasm.

The endoplasmic reticulum forms the great tubular membranous system which extends throughout the cytoplasm of cells. It may be rough-surfaced when it is studded with ribonucleoprotein (RNP) particles and termed also in this form the 'ergastoplasm' (Porter, 1953; Claude, 1955) or smooth-surfaced with no attached RNP granules or 'ribosomes'. The ergastoplasm is concerned with the synthesis of protein for extracellular secretion (Weiss, 1953; Wissig, 1963; Caro and Palade, 1964) whilst the

Organelles and inclusions

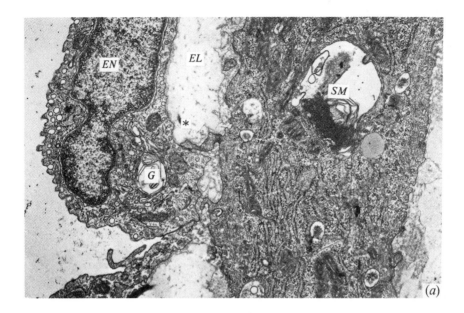

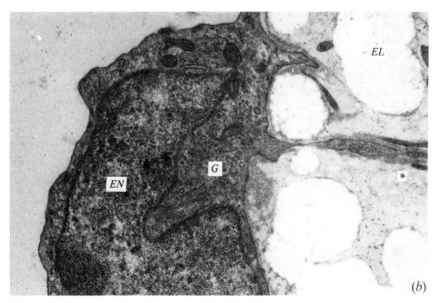

Figure 21. Two examples of aortic endothelial cells (*EN*) with well-developed Golgi regions (*G*). In both instances these cells come into close relation (∗) with cells of the media (*SM*) via gaps in the internal elastic lamina (*EL*). (*a*) Rat aorta, magnification ×9000. (*b*) Budgerigar aorta, magnification ×17500.

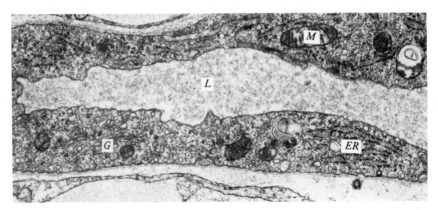

Figure 22. A small vessel of young granulation tissue in the rabbit. Plump endothelial cells enclose the lumen (*L*) and contain Golgi material (*G*), rough endoplasmic reticulum (*ER*) and mitochondria (*M*). Magnification × 17000.

smooth-surfaced form of endoplasmic reticulum is more involved in lipid metabolism (Strauss, 1963; Brenner, 1966; Goodman *et al.*, 1968). The endoplasmic reticulum of endothelial cells is usually considered to be poorly developed and to form an insignificant part of the cell as a whole (Palade, 1953a; Policard *et al.*, 1957; Fawcett, 1959; Rhodin, 1962a; French, 1963), but certainly in embryonic and regenerating endothelium it is extremely well developed (Palade and Porter, 1954; Cliff, 1963; Schoefl, 1963) (figures 22, 23, 38). When aortic endothelium is sectioned *en face* so as to allow examination of larger areas of sectioned endothelial cytoplasm more or less equivalent to the areas of cytoplasm normally seen in more conventionally shaped cells, then considerable amounts of ergastoplasmic membranes are found, most especially in the young and the very old age groups (Gerrity and Cliff, 1972).

The presence of both endoplasmic reticulum and a Golgi apparatus is morphological expression of cellular secretory activity. There are, in addition to those mentioned above in the section on Golgi apparatus, other secretory functions that endothelium may perform. During the embryonic development of blood vessels and the regeneration of blood vessels in post-natal animals, changes occur that can well be attributed to secretory activity of endothelial cells. As described by Thoma (1893, 1896), who acknowledged the similarity of his description to the methods of new vessel formation proposed by Billroth (1856), the blood vessels in their earliest embryological development consist of solid cords of endothelial cells with no evidence of a lumen. The lumen develops initially as a series of fluid-filled spaces separating the margins of the endothelial cells, which by growth and confluence eventually form the hollow lumen of the vessel. A similar process occurs in the growth of new vessels in wound

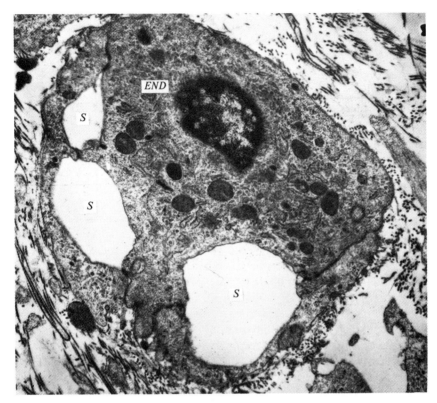

Figure 23. An endothelial sprout (*END*) from granulation tissue in a rabbit ear chamber. Separating the endothelial cells are spaces (*S*) which will coalesce to form the lumen of the new vessel. Magnification × 15 000.

healing (Cliff, 1963, 1965). I have observed the formation of intercellular spaces in endothelial sprouts invading the rabbit ear chamber with their subsequent fusion with the vessel lumen situated proximally, using time-lapse cinemicroscopy and electron microscopy (figure 23). Haar and Ackerman (1971) have similar electron microscopical evidence for inter-cellular development of spaces which coalesce to produce the vascular lumen in the development of blood vessels in the mouse yolk-sac. These appearances of spaces being formed between endothelial cells that initially were a solid cord suggest very strongly some secretory activity on the part of the endothelial cells. This idea is supported by both Lewis (1931) and Sabin (1921) who considered it very probable that the endothelial cells of developing vessels secreted the plasma that formed in the new lumen.

Hydrolysis of chylomicron triglyceride, which is an obligatory step in its passage from the blood to the tissues, appears to take place either entirely at the endothelial surface lining the vessel lumen (Schoefl and

French, 1968) or to be initiated there and continued within the endothelial vesicles (Blanchette-Mackie and Scow, 1971). This hydrolysis is produced by an enzyme known as clearing factor lipase. Its level in the blood stream rises sharply in response to heparin injection but it appears that it is produced by the somatic cells of particular tissues and not by the endothelial cells.

Finally, the consistent finding of 'reactive' endothelium in various organs undergoing the allograft rejection process is worthy of note. Such endothelial cells containing increased amounts of endoplasmic reticulum and free ribosomes have been reported by Pedersen and Morris (1970) in renal allografts and by Wiener, Lattes and Pearl (1969) in skin allografts, as well as being noted by myself in a number of different types of allograft. How this stimulation of its metabolic machinery is related to the hyperplasia of the endothelium that contributes to vascular occlusion within allografts is not clear (Hume, 1968–9; Lund and Jensen, 1970b; Heron, 1971). Presumably both are part of the very limited repertoire of responses possessed by endothelial cells to noxious stimulation.

Endothelial cells, when observed *in vitro* by phase-contrast microscopy, are seen to possess moderate numbers of filiform mitochondria (Woodard and Pomerat, 1953). In ultrathin sections examined with the electron microscope these bodies usually present relatively short oval profiles with moderate numbers of parallel transverse *cristae* (Palade, 1953b; Oberling, 1959) and intercristal dense bodies (Peachey, 1964) (figures 17, 22). The presence of such a mitochondrial population indicates that endothelial cells have only modest oxidative energy demands as these highly characteristic organelles are the sites of the oxidative phosphorylating process of cell metabolism (Schneider, 1959).

In the context of the membranous organelles of endothelium the system of surface-connected caveolae (Yamada, 1955) and vesicles (65–70 nm diameter) which arise from the luminal, abluminal and junctional aspects of the plasma membrane must also be considered (figures 24, 26). These structures are generally highly developed within endothelial cells (Moore and Ruska, 1957; Morato and Ferreira, 1957; Buck, 1958; Palade, 1960; Jennings, Marchesi and Florey, 1962; Florey, 1966; Casley-Smith, 1968; Bruns and Palade, 1968a, b; Leak, 1971) but they may also be found in comparable degrees of development in smooth muscle cells (Rhodin, 1962a; Burnstock and Merrillees, 1964) and mesothelial cells (Baradi and Hope, 1964; Fedorko and Hirsch, 1971) (figures 31, 33). There have been extensive studies performed on their morphology, interrelations and possible physiological role. In vascular permeability for non-lipid-soluble materials this system of caveolae and vesicles within endothelial cells has been studied particularly by Palade (1960) and Bruns and Palade (1968a, b), as well as by others (Jennings, Marchesi and Florey, 1962; Pappas and Tennyson, 1962; Jennings and Florey, 1967; Hüttner, More and Roma,

Organelles and inclusions 49

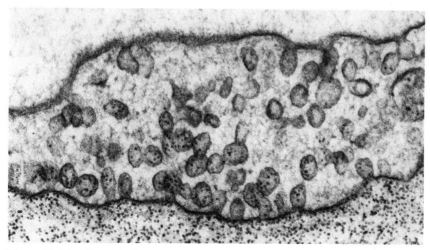

Figure 24. Blood vessel of lactating murine mammary gland one hour after injection of ferritin solution over the gland: there are numerous ferritin particles related to the basal surface of the endothelium and many have entered endothelial vesicles from this surface. Magnification × 99 000. Electron micrograph reproduced through the kindness of Dr G. I. Schoefl.

1970). The findings of all these investigators indicate that this membranous system is involved in the transport of test materials from the blood stream across the endothelial barrier to enter the tissues. Such conclusions are based on electron microscopic studies of the passage of various tracers such as ferritin (Farrant, 1954), which must be purified of the cadmium often used in its preparation (figure 24), and various colloidal preparations of carbon, gold, thorium dioxide and mercuric sulphide, which all have atoms with high enough atomic numbers within their molecules to render them electron-dense in the electron microscope and which have sizes more or less comparable to those of proteins (see Karnovsky, 1968). Ferritin alone in this group can be considered a biological substance, being an iron-storing protein obtained usually from the spleens of horses. The other approach to this problem of obtaining suitable electron-dense tracers which simulate the movements of proteins from the blood stream to the tissues is to use enzymes such as horseradish peroxidase (HRPO) (Graham and Karnovsky, 1966) which has a molecular weight of 40 000. This plant enzyme can be localized in tissue that has been fixed with glutaraldehyde (Sabatini, Bensch and Barrnett, 1963) by reacting it with hydrogen peroxide in the presence of 3,3'-diamino-benzidine. The reaction product so produced becomes a slightly flocculent electron-dense deposit on post-fixation with osmium tetroxide. It is unfortunate that this plant protein has definitely been shown to cause release of histamine and serotonin from mast cells (Smith, 1963) in both the rat and guinea pig, leading to increased vascular leakage (Cotran and Karnovsky, 1967). The fact that pharmacologically active substances such as these have been implicated in conditions of pathologically increased vascular permeability (see pp. 153–4) renders very difficult the interpretation of the results obtained with HRPO in relation to normal physiological permeability. Cotran and Karnovsky established that the mouse does not have this reaction to HRPO. Gerrity, working on rats in my laboratory, has shown that a 50% fall in blood pressure occurs acutely after intravenous injection of HRPO as a tracer. It should also be borne in mind that there are endogenous peroxidases present within a variety of cells, which includes those of the myeloid series (Graham and Karnovsky, 1966; Bainton and Farquhar, 1970), peritoneal macrophages and Kuppfer cells (Widmann, Cotran and Fahimi, 1972) and thyroid epithelial cells (Strum and Karnovsky, 1970). In addition haemoglobin has peroxidase activity (Graham and Karnovsky, 1966; Goldfischer et al., 1970), but to date this readily available protein has not been used in the electron microscopic study of vascular permeability as such, except fortuitously by Latta (1970) in the renal glomerulus where it was visualized by virtue of its own electron-density and not through its enzymatic activity. Majno (1965) has discussed very fully the problems attendant on the use of the various electron microscopic labels employed for studying vascular permeability and emphasizes the difficulty of interpreting the results obtained with them.

Bearing this in mind it is found quite consistently that ferritin molecules which have diameters of 10 nm (Farrant, 1954) are taken up by the endothelial luminal caveolae and vesicles and apparently traverse the endothelial cytoplasm within the vesicles which then fuse with the abluminal plasma membrane to discharge the ferritin into the extra-endothelial space (Jennings and Florey, 1967; Bruns and Palade, 1968b; Hüttner et al., 1970). Particles of colloidal carbon, which are considerably larger, also seem to traverse endothelial cells by a similar pinocytic method (Mikata and Niki, 1971). In the cases of other tracers, however, endothelial permeability has been demonstrated simultaneously at the endothelial cell junctions and, apparently, within the vesicular system, as for example with HRPO (Karnovsky, 1967; Hüttner et al., 1970) and colloidal preparations of thorium dioxide, gold and saccharated iron oxide (Pappas and Tennyson, 1962; Jennings and Florey, 1967). The function of cellular vesicles in the carriage of materials across a cell rather than into the cell for its own use has been termed 'cytopempsis' (Moore and Ruska, 1957).

Estimates have been made of the amount of time such vesicles spend in each of the three phases involved in their postulated transendothelial carriage of non-lipid-soluble materials. The phases are: that of initial continuity with the luminal plasma membrane, when they are available to take up materials present in the vessel lumen; a transit phase, when the vesicles in some way have severed connection with the luminal cell surface and are free to move within the endothelial cytoplasm; and, finally, a phase of abluminal surface-connection when the contents of the vesicles are able to be discharged into the extravascular space. Bruns and Palade (1968a, b), in an extremely painstaking study involving three-dimensional models constructed from serial electron micrographs of endothelial cells, concluded that the vesicles spent 30% of their time in continuity with the luminal surface, 30% of their time in transit and 40% of their time in continuity with the abluminal endothelial surface. The absolute transit time they estimated as between 24 and 34 seconds. Shea, Karnovsky and Bossert (1969) made use of the results of Bruns and Palade for computer models of a hypothetical system, and by making various assumptions, including a figure for endothelial cytoplasmic viscosity, arrived at a transit time for vesicles of 1.09 seconds. Casley-Smith (1968) in a theoretical approach to vesicular transport proposed attachment times to membranes of about 2 seconds and a transit time of about 1.5 seconds. It is implicit in all these considerations that vesicles are continuously shuttling to and fro bearing their freight rather in the manner of a bucket-chain or a game of ping-pong with the vesicles persisting to take up and discharge materials more or less indefinitely and not being formed on one plasma membrane to 'die' by being absorbed into the other plasma membrane following fusion. This concept is supported by the work of Jennings and Florey (1967) who could find no evidence of metabolic energy needs for vesicle labelling or transport across endothelium and who concluded that the vesicles were quite long-lived structures within the cells. It has been proposed that the actual movement of free vesicles within the cytoplasm is due to Brownian motion produced by the thermal energy of the system (Shea and Karnovsky, 1966; Casley-Smith, 1968; Shea, Karnovsky and Bossert, 1969). Shea and Karnovsky stressed the importance of an initial energy dependent intrusion of the surface-connected vesicles by their neck regions in order to get them started on their journey. They calculated that in a truly random form of motion, such as occurs due to thermal agitation, the probability of a vesicle moving across to make connection with the opposite cell surface rather than fusing with the surface it has left is x/y, where x is the distance the vesicle has been intruded by its neck from the surface and y is the distance separating it from the opposing cell surface. Such considerations are purely theoretical and, as pointed out by Bruns and Palade (1968b), no experimental estimates of the types of forces or sources of energy involved in vesicular transport are feasible at this time. In connection with the random to and fro shuttling of vesicles it is interesting to note that there is some evidence for the reverse movement of tracers occurring from the abluminal to the luminal surfaces of endothelium, as would be expected from such a postulated model. Mikata and Niki (1971) found that thorium dioxide particles occasionally moved in the reverse direction across the endothelium of post-capillary venules. Schoefl

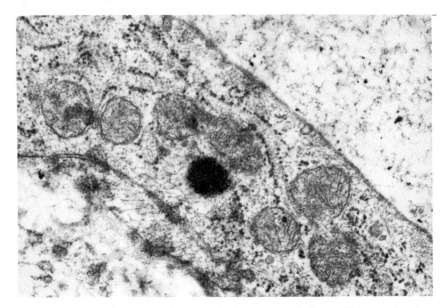

Figure 25. The endothelium of a recently formed venule in a rabbit ear chamber with a large spherical vesicle (near middle of figure) containing electron-dense particles of thorium dioxide following their intravenous injection. Scattered particles of thorium dioxide are present in the vessel lumen at the upper right of the figure. Magnification × 45 000.

(personal communication) has detected ferritin particles sometimes moving in the reverse direction across the endothelium of the lactating mammary gland (figure 24).

A very real complication in all considerations of uptake of tracers by endothelial cells is the definite phagocytic power possessed by these cells (figure 25) in normal vessels of all sizes (Buck, 1958; Cotran, 1965; Schoefl and French, 1968; Mikata and Niki, 1971), in vessels involved in inflammation (Majno and Palade, 1961; Cotran, Guttuta and Majno, 1965) and in wound healing (Foot, 1921; Cliff, 1963), in aging arteries (Gerrity and Cliff, 1972) and in hypertensive (Hüttner et al., 1970) and atherosclerotic arteries (Woerner, 1951; Veress et al., 1970). This generalized phagocytic power of endothelium is considered to be more highly developed in embryonic blood vessels than in mature vessels (Beard and Beard, 1927). The process of phagocytosis implies the presence within the cytoplasm of lysosomal bodies which contain within a membrane-bound organelle a variety of hydrolytic enzymes with acid pH optima (Duve, 1959; Novikoff, 1961; Weissman, 1965a, b). Recently Ts'ao (1970) has been able to demonstrate the presence of lysosomes within endothelial cells of both the aorta and the inferior vena cava of normal rats and rabbits using a modified histochemical technique for demonstrating acid phosphatase activity with the electron microscope. It has previously been shown that endothelial cells of damaged vessels have high levels of acid phosphatase activity (Lesko, Babala and Lojda, 1966), being interpreted as the usual lysosomal response of degenerating cells (Novikoff, 1963; Weissman, 1965a, b). The possible role of blood platelets in supplying the lysosomal enzymes of endothelium has been discussed above. A distinction must be made between this acid phosphatase activity of endothelium and the more generalized alkaline phosphatase activity of these cells which distinguishes them from lymphatic endothelium (Kabat and Furth, 1941) and has been used in the identification of blood capillaries both in normal and pathological tissues (Fanger and Barker, 1960) and in tissue culture (Farnes and Barker, 1963).

It would be a fair summary of current experimentally-based conclusions to say that the membranous vesicles are involved to some degree in the transport of non-lipid-soluble materials across the endothelial membrane.

There is a second system of vesicles which has been described in endothelial cells by Stehbens (1965). The vesicles are about the same general size as the endothelial vesicles of Palade (1953a) but are distinguished from these by having thicker, more electron-dense walls (10–30 nm across) with fine projections or bristles extending from their outer surfaces. These bristles have led to these structures being called 'acanthosomes' by some authors. Stehbens (1965) has suggested that these endothelial structures may be involved in protein transport and was able to demonstrate thick-walled caveolae in continuity with the plasma membrane. Acanthosomes have been shown to be associated with protein secretion (Fahrenbach, Sandberg and Cleary, 1966) and uptake (Rosenbluth and Wissig, 1964) in cells other than endothelium, but recently Joó (1971) has made observations linking enhanced permeability to macromolecular substances with increased numbers of coated vesicles in brain capillaries.

Microfibrils and microtubules, organelles found in endothelium, as in many other cell types, have already been discussed (p. 43).

PORES, PERMEABILITY AND FENESTRAE

Various attempts have been made to equate the vesicular transport system with the 'large pores' of certain physiological models of capillary permeability which have been proposed to account for the observed restricted passage of non-lipid-soluble materials and water (Mayerson et al., 1960). Bruns and Palade (1968b) considered that the endothelial vesicles corresponded to the 'large pores' but could not identify the system of 'small pores' also required by the model. Hüttner and others (1970) equated the vesicles of endothelial cells with the 'large pores' and their cell junctions with the 'small pores', as was also proposed with rather more reservations by Karnovsky (1968). It must be appreciated that the physiological model involving small and large pores is only one of several models of capillary permeability. Pappenheimer, Renkin and Borrero (1951) proposed a model to account for their observations on muscle capillary permeability which required the existence of a system of water-filled pores occupying about 0.2% of the capillary surface area which could either be uniform pores of 3.0 nm radius or rectangular slits with widths of 3.7 nm, or else pores with a Gaussian distribution of radii of 2.4 ± 1.2 nm (s.d.). These figures have been modified in a more recent publication by Landis and Pappenheimer (1963) who gave figures, for muscle capillaries once again, in which the area occupied by the pores was less than 0.1% of the capillary surface area and their radius was 4.0 to 4.5 nm. In addition to these 'small' pores they postulated the existence of a much smaller number of 'large pores' or 'leaks' probably located in the venules themselves. This model of capillary permeability is in accord with that deduced by Grotte (1956) who concluded that his observations were explicable in terms of an isoporous membrane with pore radius of 3.5–4.5 nm with a smaller number of 'leaks' with radii ranging from 12.0 to 34 nm, which, as their name implies, allow bulk flow of water to occur, as opposed to the diffusion that is postulated to occur through the various 'pores' of these models. The ratio of 'leaks' to 'pores' may vary a hundred-fold depending on the site within the body, since Grotte gives this ratio for the hind-leg preparation as 1/34000 and for the liver as 1/340.

It has been mentioned above that the small pores may correspond to the endothelial cell junctions. Landis and Pappenheimer (1963) considered that their postulated pores would conform well with the cellular junctions of the endothelial membrane both in terms of the proportion of area they occupy in the membrane and in their absolute dimensions, with 10 nm slits separating the related plasma membranes. There was some dismay when blood vessel endothelial junctions were found to have regions of membrane fusion, as described above. In the case of the blood vessels of the brain these regions of membrane fusion have been shown to be impermeable to tracers such as HRPO in normal animals (Bodenheimer and Brightman, 1968; Giacomelli, Wiener and Spiro, 1970) but they may become permeable to this tracer in hypertensive animals. Karnovsky has shown that in capillaries other than those of the brain HRPO passes down the endothelial junctions since the regions of membrane fusion, or zonulae occludentes, are not continuous bands but are circumscribed areas forming 'buttons' of attachment (Karnovsky, 1967). The zonulae occludentes frequently have membrane separations of 4.0 nm which are permeable to HRPO and lanthanum (Karnovsky, 1968). Majno (1965) has stated that the regions of the endothelial junctions should be considered as filters rather than seals. Bruns and Palade (1968a, b) speculated as to whether the various procedures involved in the fixation and preparation of tissues for electron microscopic examination could alter the cell junctions of the endothelium so as to reduce the gaps between the cells from a 7.0–9.0 nm separation *in vivo*, or else whether in fact the 'pores' of physiological models may have been formed by continuous chains of interconnecting vesicles extending across the endothelial cell to link its two surfaces. Such 'chains' would have necks 9.0 nm in width where they arose from the plasma membrane and Bruns and Palade considered the possibility that such structures existing *in vivo* might be destroyed at some stage during fixation and processing. A somewhat similar concept of vesicle 'chains' is supported by Jennings and Florey (1967) who postulated that channels linking the two endothelial surfaces may be formed by strings of vesicles randomly connecting and breaking apart again.

I think some consideration should be given here to the difficulties that attend the interpretation of electron micrographs of biological materials in relating them to the dynamic reality of living systems. It is very difficult in the face of the often beautifully intricate and orderly arrangements made visible by correctly applied techniques of electron microscopy to realize that such images are not of the living tissue itself but of an artefact. In fact the final judgment as to whether an electron micrograph of biological tissue adequately represents the true ultrastructural organization must be subjective. This applies in the evaluation of the relative merits of different fixatives (Palade, 1952; Porter and Kallman, 1953; Bahr, 1954; Barer and Meek, 1960; Sabatini *et al.*, 1963; Bloom, 1970; Wisse, 1970; Peracchia and Mittler, 1972), the buffers in which they are used (Bennett and Luft, 1959; Wood and Luft, 1965), the osmolarity (Caulfield, 1957; Maunsbach, 1966) and ionic composition and dielectric constant of its solution (Robbins, 1961; Cooke and Fay, 1972). Similar subjective estimates of the value of employing various plastics for embedding tissue for ultrathin sectioning (Luft, 1961) and the merits of various forms of heavy metal staining (Watson, 1958a, b) must also be made.

With this in mind let us return to the ultrastructural appearance of the endothelial cell junction. One feature that the zonula occludens region has is a look of permanence. However Farquhar and Palade (1963) have found that regions of membrane fusion that produce zonula occludens types of structures can appear between red blood cells and endothelial cells in both the rat and the guinea pig, indicating that such structures may be quite labile. Curtis (1964) measured the distance separating the cell surface and the substrate at sites of cellular attachment in tissue cultures, using an optical interference method, and detected a cycle of changes occurring during processing for electron microscopy. He concluded, however, that the contact regions that were finally obtained, with separations of 10–20 nm, probably reflected quite accurately the situation that existed before fixation was commenced. Goodenough and Revel (1970) showed that the successful demonstration of gap junctions in aldehyde-fixed tissues, with 2.0 nm separation of the opposing plasma membranes of epithelial cells, depended upon the method of post-fixation and processing employed. This indicates that 'fixed' tissues are certainly not incapable of changing their structural interrelations in the course of processing for electron microscopy. For this reason it is essential that in evaluating the true ultrastructure of cells or tissues results from a variety of different techniques should be employed. The technique of freeze–fracture or freeze–etch preparation of biological material for electron microscopy (Staehelin, 1968) is of great value as an alternative method of preparing tissues for examination by electron microscopy as it eliminates the need for chemical fixation, dehydration and embedding. This technique has been applied recently to the study of endothelium (Leak, 1971; Leeson, 1971) and has confirmed the existence of the endothelial vesicles and so renders the possibility of these structures being formed from pre-existing tubular structures in the course of conventional fixation and processing for electron microscopy (Hodge, 1956, and discussed by Jennings and Florey, 1967) most unlikely. This technique has not yet thrown any light on the nature of the endothelial junctions, but must obviously do so in the future.

In addition to the vesicular system and the endothelial cellular junction regions there is a third feature of endothelial cells that must be considered in relation to vascular permeability. These are the endothelial *fenestrae* (L. = windows) which are definite small holes that pierce certain endothelial cells from luminal to abluminal surface and whose margins are completely lined by the endothelial plasma membrane (figures 11, 26, 46). These holes are generally circular but they may apparently also be oval in shape, as described by Wisse (1970) in the liver. Their diameters vary considerably depending upon their site within the body but within any particular region they are fairly uniform in size. Hence the diameters of fenestrae present in endothelial cells of the choroid plexus and the ciliary body of the eye are given as 30–40 nm (Pappas and Tennyson, 1962), for the choriocapillaris of the eye as 40–50 nm (Leeson, 1971), for endocrine glands, gut, pancreas and kidney (excluding glomerulus) as 30–50 nm (Fawcett, 1963), whilst figures of 100 nm are given for both hepatic (Wisse, 1970) and renal glomerular (Fawcett, 1963; Latta, 1970) fenestrae. The fenestrae in fact establish fine channels directly linking the vessel lumen with the extra-endothelial space. They are generally found to be closed by a fine moderately electron-dense membranous condensation, or diaphragm, 4.0 nm (Karnovsky, 1968) to 6.0 nm (Rhodin, 1962*b*) in thickness in which a central thickening or 'knob' 10 nm in diameter may often be seen. The material forming the diaphragm can be traced into continuity with the outer dark leaflet of the trilaminar unit membrane (Robertson, 1959) in favourable sections and is considered to be probably of polysaccharide or mucoprotein composition (Luft, 1965; Karnovsky, 1968). Certain investigators have found fenestrae to have closing diaphragms either very infrequently (Latta, 1970) or not at all (Wisse, 1970; Vegge, 1971). It would be unwise to attach any great importance to such a feature in view of my own experience with the fenestrae of renal blood vessels, which have diaphragms readily identifiable in tissue fixed by immersion but absent in tissue fixed by vascular perfusion with the same fixative solution. It would appear that unless all fixation and processing procedures are identical the presence or absence of closing diaphragms cannot be interpreted as having any functional significance.

Fenestrae may occupy large areas of certain endothelial membranes; Webber and Blackbourn (1970) estimated that up to 30% of the area of the glomerular endothelium could be taken up by fenestrae. Leeson (1971) has shown in some beautiful freeze-etch preparations of the choriocapillaris of the eye that regular spacing of the fenestrae frequently occurs within the endothelial cells and that regions where fenestrae are completely absent may lie between regions of fenestrated cytoplasm in the same endothelial cell. Fenestrae may occur in flaps of endothelial cytoplasm that extend into the vessel lumen, frequently in relation to endothelial cell junctions (Stehbens, 1965; Ludatscher and Stehbens, 1968) (figure 26). It is hard to imagine that fenestrae in such situations as this could have any functional role, but it does illustrate the intrinsic tendency of certain endothelia to produce these structures. There are now in the literature numerous reports of the presence of diaphragms apparently very similar to those of the fenestrae which close the mouths of endothelial caveolae arising from both the luminal and abluminal endothelial surfaces (Karrer, 1960a; Bruns and Palade, 1968a; Stehbens and Ludatscher, 1968; Takada and Gore, 1968; Takada, 1970) (figure 26). It is found that fenestrae only occur in the thinnest regions of the endothelial membranes of the microcirculatory vessels, most frequently identified as capillaries (Palay and Karlin, 1959; Karrer, 1960a; Pappas and Tennyson, 1962; Rhodin, 1962b; Zelander, Ekholm and Edlund, 1962; Fawcett, 1963; Ludatscher and Stehbens, 1968) but at times more specifically as venous capillaries (Rhodin, 1968) or as the transitional regions between capillaries and venules (Takada, 1970) or, occasionally, as venules (Stehbens, 1965; Takada, 1970). It is, in fact, in these regions that the thickness of the endothelial cytoplasm becomes comparable to, or less than the average diameter (65-70 nm) of the endothelial vesicles. This, of course, excludes the neck region of a caveola which intrudes into the cytoplasm. If this region is also included, then it becomes even easier to visualize the situation wherein a caveola still attached to one surface of an endothelial cell could merge with the plasma membrane on the opposite surface of the cell (Karrer, 1960a). In this way a hole with a unit membrane wall would be produced linking the two surfaces of the cell. The closing

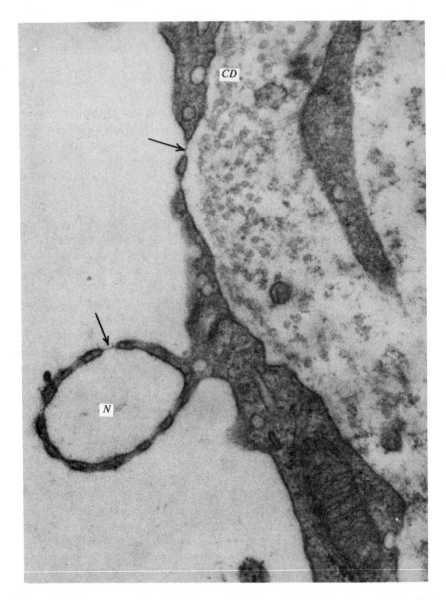

Figure 26. A human renal peritubular vessel with endothelial fenestrae with closing diaphragms (arrows). Some fenestrae link the vessel lumen and the extravascular space but others are within an endothelial flap that forms a fenestrated intravascular 'necklace' (*N*). A vesicle is closed by a diaphragm on the basal endothelial surface (*CD*). Kidney removed at operation for chronic renal failure. Magnification × 75 000.

diaphragms, when present, could then be derived from the closing diaphragms of the mouths of caveolae of endothelial cells, as discussed above. In fact, endothelial vaculoes simultaneously connecting the two cell surfaces can also be identified (Blanchette-Mackie and Scow, 1971). It is possible to envisage an energy dependent system under the metabolic control of the endothelial cell which controls the formation and maintenance of a system of fenestrae. The extremely ephemeral nature of the fenestrated 'sieve-plates' lining the hepatic sinusoids has been demonstrated by Wisse (1970). Karrer considered that the fenestrae he observed were probably very labile transient structures related to the breaking through of vesicles.

There is no doubt that the presence of fenestrae in certain sites greatly enhances the permeability of the endothelium to water and contained solutes, including proteins. This is well illustrated in the fenestrated endothelium of the renal glomerulus which has been shown to allow rapid passage via the fenestrae of proteins such as HRPO (Graham and Karnovsky, 1966; Karnovsky, 1968; Webber and Blackbourn, 1970), haemoglobin (Latta, 1970), ferritin (Farquhar and Palade, 1960; Webber and Blackbourn, 1970) and human myeloperoxidase (mol. wt. 160000–180000) (Graham and Karnovsky, 1966). The increasing permeability of the mammalian glomerulus that develops with maturity has been linked to the increasing numbers of fenestrae within the endothelial cells (Webber and Blackbourn, 1970). The main filtration barrier within the glomerulus has been localized to being either at the glomerular basement membrane (Farquhar, 1960; Webber and Blackbourn, 1970) or at the slits between the epithelial foot processes of the glomerulus (Graham and Karnovsky, 1966; Latta, 1970). Even so, proteins such as HRPO and ferritin do reach the urinary space (Graham and Karnovsky, 1966; Webber and Blackbourn, 1970). The renal tubular epithelium shows evidence of HRPO uptake from the glomerular filtrate (Graham and Karnovsky, 1966). This, taken in conjunction with the findings of Bourdeau, Carone and Ganote (1972), which indicate that a gradient of uptake of albumin occurs normally in the proximal renal tubule epithelium, points to the fact that the glomerulus does not necessarily produce a true ultrafiltrate from the plasma.

The fenestrated endothelium of the liver sinusoids (Laschi and Casanova, 1969) is likewise associated with very high permeability of these vessels to water and solutes up to 250000 mol. wt. (Greenway and Stark, 1971) or even 412000 mol. wt. (Mayerson *et al.*, 1960). In addition, the fenestrae of these vessels may well play a role in allowing the selective passage of blood chylomicrons (Wisse, 1970) and also β-lipoproteins or VLDL fraction into the liver (Grubb and Jones, 1971). Fenestrae have been seen allowing direct access to the blood vessel interior of microvilli projecting from hepatocytes (Wisse, 1970), and also apparently the same phenomenon has been identified in the adrenal cortex where a shed microvillus penetrated the diaphragm of an endothelial fenestra to enter the blood stream (Brenner, 1966). Pappas and Tennyson (1962) investigated the properties of the blood vessels of the choroid plexus and the ciliary body of the eye. These vessels shared the features of very thin endothelial membranes with fenestrae present with those from other regions which are known to have high levels of water transport, such as the intestinal villi (Palay and Karlin, 1959) and the renal tubular capillaries (Rhodin, 1962b). However, they found no evidence for the passage of particles of thorium dioxide, saccharated iron oxide or gold-sol through the fenestrae of these vessels. What passage did occur was in the thicker regions of the endothelial membranes via the vesicles and the intercellular junctions. Vegge (1971) using HRPO, however, has demonstrated very ready passage of this material through the fenestrae of the blood vessels of the ciliary process of the eye and, by way of similarity with the renal glomerulus, found that the main barrier responsible for maintaining the extremely low protein content of the aqueous humour (5–15 mg per 100 ml in man) was the junctional regions of the related epithelial cells.

It is very probable that endothelial fenestrae are associated with high levels of net fluid uptake into vessels in certain sites, such as in the intestinal villi and in the counter-current concentrating systems in the vasculature of the kidney (Berliner *et al.*, 1958; Longley, Banfield and Brindley, 1959; Moffat, 1967) and in the rete mirabile of the teleost swim bladder (Fawcett, 1963). The evidence obtained by Milhorat *et al.* (1970) that indicated that albumin injected into the cerebral ventricles of human hydrocephalics was absorbed into the blood vessels of the choroid plexus may also be evidence for the absorptive activity of the fenestrated endothelial cells in these situations. There remain however, as pointed out by Florey (1966), a variety of sites where fenestrated endothelium occurs with no definite physiological properties being ascribed to it. Such sites include muscle fascia (Rhodin, 1968), both muscle (Bruns and Palade, 1968a) and lamina propria (Takada and Gore, 1968) of the tongue, salivary (Takada, 1970) and pancreatic exocrine (Zelander *et al.*, 1962) glands, the median eminence of the pituitary (Mikami *et al.*, 1970) and the carotid body (Böck, Stockinger and Vyslonzil, 1970). It has also been suggested that the endothelial fenestrae might correspond to the large pores postulated in physiological models for capillary permeability (Fawcett, 1963; Bruns and Palade, 1968b). The presence of such structures in the venous ends of capillaries and in the small venules, pointed out earlier in this chapter, may well be correlated with the in-vivo observations of Landis (1964) who detected considerably greater permeability for colloid in such regions of the microcirculation.

Consideration has so far only been given to the movement of non-lipid-soluble materials across the endothelium. Lipid-soluble materials such as urethane, paraldehyde, triacetin and the respiratory gases, oxygen and carbon dioxide, are all considered to pass directly through the endothelial membranes (Pappenheimer et al., 1951; Landis and Pappenheimer, 1963). Their movement is very rapid and as shown in the case of various glycerol–acetic acid esters the rate of movement is directly proportional to their oil–water partition coefficient. The very extensive system of endothelial caveolae and vesicles which have been shown by Karnovsky (1968) to be practically all surface-connected and which effectively increase the membranous surface area of the endothelial cell by an additional 200% (Bruns and Palade, 1968a) may well have an important function in increasing the membrane area available for diffusion of lipid-soluble substances, most particularly the respiratory gases oxygen and carbon dioxide.

NUCLEUS AND MITOSIS

Each endothelial cell possesses a single centrally situated nucleus which is flattened in the same plane as and elongated in the long axis of the cell. The endothelial nucleus, in common with the nuclei of other cells, contains nucleolar material and is enclosed within a double membrane system, or envelope, in which nuclear pores can be identified. Nuclear orientation with respect to the distribution of their long axes within blood vessel walls is used at times to help distinguish arterial, capillary and venous types of vessels in sections. Very characteristic nuclear deformations have been observed within the endothelial cells of minute vessels rendered highly permeable by treatment with mediators such as histamine, serotin and bradykinin (Majno, Shea and Leventhal, 1969) (figure 27). Majno and coworkers took these changes as evidence of endothelial contraction which pulled apart the venular endothelial cells at their junctions and resulted in the formation of endothelial intercellular gaps so characteristic of these vessels in areas of inflammation (Marchesi, 1962; Movat and Fernando, 1963; Schoefl, 1963; Cotran, 1967).

Sex chromatin markers, or Barr bodies (Barr, Bertram and Lindsay, 1950) may be identified within endothelial nuclei and have been used to help elucidate the re-endothelialization of certain tissue allografts when donor and recipient are of opposite sexes (Williams et al., 1971).

Endothelial nuclei are capable of karyokinesis as seen in mitotic divisions during embryonic blood vessel development (Thoma, 1893; Kampmeier and Birch, 1927) and post-natal growth and maturation of blood vessels (Gerrity and Cliff, 1972). There is no doubt that aortic endothelial cells have well-developed proliferative capacities when suitably stimulated, as shown in the identification of numerous endothelial mitoses associated with the repair of defects produced by mechanical damage to the aortic endothelial membrane (Poole et al., 1958; Bondjers

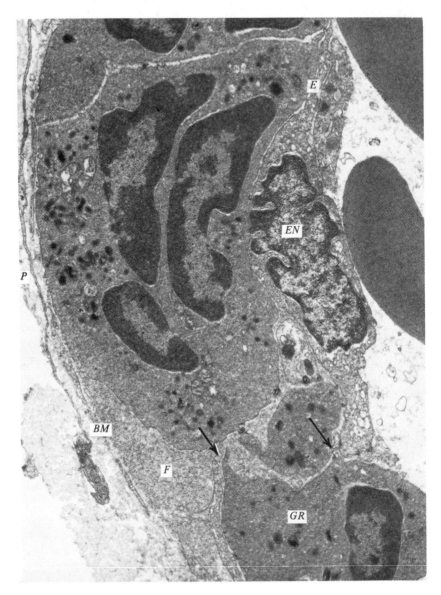

Figure 27. A rat venule fixed 5 minutes after topical application of histamine 1/1000. Granulocytes (*GR*) are migrating through a large endothelial gap (between arrows) to lie between the endothelium (*E*) and the basement membrane (*BM*). Increased vascular permeability is shown by the accumulation of intravenously injected iron label (*F*) between the endothelium and the basement membrane. The endothelial nucleus (*EN*) has both folds and pinches in its margin. Tenuous pericyte processes (*P*) surround the vessel. Magnification ×17500.

and Björnheden, 1970). Prosthetic grafts into aortae of various animals formed new endothelial linings within a matter of weeks (Florey et al., 1962; Jordan et al., 1962; Warren and Brock, 1964). The new endothelium spread over the plastic grafts from each end and also from islands of endothelium that developed separately towards the centres of the grafts. These islands are probably related to the ingrowth of *vasa vasorum* (see p. 125) observed passing through the walls of knitted prostheses by Warren and Brock (1964). The formation of endothelial membranes that covered plastic hubs, suspended by fine sutures in the centres of dacron aortic prostheses (Stump et al., 1963; O'Neal et al., 1964), does however demonstrate that direct seeding of endothelium can occur, presumably analogous to the seeding of fibroblastic cells observed by Ross and Lillywhite (1965) associated with damage to blood vessel walls. In addition Williams et al. (1971) demonstrated that islands of host endothelial cells formed a new lining that replaced the shed donor endothelium of aortic allografts. Initial results indicate that such seeding may be by bone marrow derived cells. Mitotic proliferation of endothelium has been observed in tissue cultures of aortic explants (Kasai, Pollak and Nagasawa, 1964) bone marrow (Woodard and Pomerat, 1953) and umbilical veins (Gimbrone et al., 1974).

In the rat the endothelium of the aorta showed no change from its initial low level of [^{3}H]thymidine labelling in response to experimentally induced hypertension but that of arteries and arterioles did show an increase (Crane and Dutta, 1963). In addition, the single kidney remaining after the unilateral nephrectomy performed in these experiments as part of the measures to produce hypertension showed considerable compensatory hyperplasia and had heavy labelling of glomerular and general renal blood vessel endothelium. Vascular hyperplasia associated with the development of coronary collateral circulation showed high levels of [^{3}H]thymidine labelling and mitotic activity of endothelial cells, with the smallest vessels showing the greatest responses (Schaper, 1971).

Endothelial mitoses occur in the new vessels formed in wound healing and in granulation tissue (Thoma, 1896; Foot, 1921; Clark, 1936; Dustin and Chodkowski, 1938; Clark and Clark, 1939; Chalkey, Algire and Morris, 1946; Cliff, 1963, 1965). The stimulus that is responsible for this endothelial proliferation and new vessel formation has been suggested as being increased blood pressure (Thoma, 1893) or, closely allied to this, the conclusion of Hughes and Dann (1941), the richer blood supply at the viable margins of the lesions. This is not far removed from my own conviction that acute inflammation is an indispensible accompaniment of the repair process, as also stated by other investigators (Travers, 1844; Clark, 1936, 1946; Edwards and Dunphy, 1958). On the other hand, what may be considered as chemical forms of stimuli have also been suggested as being responsible for initiating endothelial proliferation. These include the conclusion by Williams (1959) that hypoxia, of sufficient degree and duration, was a major stimulus for endothelial ingrowth, whilst the importance of tissue extracts in promoting new vessel formation has been illustrated both for extracts of normal (Edwards, Sarmeta and Hass, 1960) and of neoplastic tissues (Cavallo et al., 1972). Also falling within this chemical group is the suggestion by Clark (1946), that the products of granulocyte degeneration and macrophage activity had growth-promoting properties in repair processes, and by Foot (1921), that the exudate developing as a result of the initial injury induced embryonic characteristics to appear in the cells involved in the healing of wounds. Clark and Clark (1939) considered that the rate of exchange of materials across the blood vessel walls probably played a role in determining the growth or regression of capillaries. White (1954) demonstrated that in-vitro outgrowth of blood capillaries occurred most prolifically at pH 7.0, or higher, and less prolifically at lower pH values. However, she concluded that the major factor for stimulating capillary growth was the initial outgrowth of fibroblasts from the explant. The suggestion made by Crocker, Murad and Geer (1970) that pericytes may exercise some form of contact inhibition to control the proliferation of blood capillaries in wound healing raises the interesting possibility that the initiating event in the endothelial proliferation and invasion associated with wound healing and with the formation of granulation tissues may be the loosening of pericyte–endothelial intercellular contacts, with migration away from the capillary walls of the pericytes at the margins of the wound.

Proliferation of endothelial cells in the lining of blood vessels of allografted organs is frequently encountered. In certain instances this endothelial proliferation has been recorded as occurring in conjunction with mesangial cell hyperplasia within the glomeruli of renal allografts (Lindquist, Guttman and Merrill, 1968) and in the glomeruli and other vessels of the kidney (Lund and Myhre Jensen, 1970*b*) but has had no crucial importance placed upon it in terms of the ultimate rejection of the allografts. Other investigations have revealed similar endothelial proliferation within the arterioles of allografted hearts (Chiba et al., 1962; Heron, 1971). From studying unmodified rejection of renal allografts in dogs Horowitz et al. (1965) considered that the allografted endothelium was stimulated by local injury to proliferate which resulted in thrombosis, vascular occlusion and tissue necrosis. Hume (1968–9), in reviewing human renal allografting, considered that in the chronic humoral form of rejection focal ischaemic necrosis in the grafts was produced by striking endothelial overgrowth which occluded the lumina of small blood vessels and which occurred as a result of antibodies directly 'attacking' the endothelial cells.

Quick (1957) drew attention to endothelial proliferation in the walls of blood vessels apparently uninvolved by thrombosis in patients with thrombotic thrombocytopaenic purpura and considered that thrombi occurred in this syndrome secondarily to changes in the blood vessel walls. It is not impossible that the proliferative response of the endothelial cells found in this syndrome may be a similar form of response to that occurring in the rejecting allograft endothelium (p. 64) particularly in view of the suggestion made by Adelson and others (1954) that in thrombocytopaenic purpura, to which they prefixed 'thrombohemolytic' in the syndrome, there is a hypersensitivity state involving red blood cell, platelet, megakaryocyte and vessel wall.

The proliferative capacity of endothelium also finds abnormal expression in the various tumours of endothelial origin. The majority of such tumours are developmental in origin, being either congenital or associated with aging (Barnard and Robb-Smith, 1945; Quick, 1957). These are the capillary haemangioma, or telangiectasis, commonly occurring as the familiar birth-mark in the skin, and the cavernous haemangioma which occurs most commonly in the liver. These tumours are benign in that they do not form metastases but they can be the cause of fatal haemorrhage, most often with those in the renal pelvis (Barnard and Robb-Smith, 1945), but Quick (1957) has also described the death of a woman due to haemorrhage resulting from combing her hair and damaging a scalp telangiectasis, the faulty haemostasis not being due to any defect in blood coagulation but to the abnormal vessel walls. In the case of senile angiomata (Campbell de Morgan spots) electron microscopy has shown that the masses of thin-walled dilated blood vessels are unusual in having numerous fenestrae present in the endothelium and very thick multi-layered basement membrane systems (Ludatscher and Stehbens, 1968; Stehbens and Ludatscher, 1968).

The *haemangioma simplex* appears to form a bridge between these benign endotheliomata and the malignant haemangioendotheliomata, in that it is characterized by a remarkable tendency to infiltration of surrounding tissues and is reported occasionally to give rise to metastases. In the true malignant haemangioendothelioma the growing edges have 'branching syncytia' which bear a striking resemblance to vasoformative cells (Barnard and Robb-Smith, 1945), as found for instance in wound healing (Cliff, 1963, 1965). This similarity extends down to the fine-structural level, where, as in the case of regenerating vessels, no 'syncytia' of endothelial cells are found and numerous intercellular spaces separate the endothelial cells (Toth and Wilson, 1971). Another common feature shared by both normally proliferating and neoplastic endothelial cells is the presence of very well developed fibroblast-like endoplasmic reticulum (Cliff, 1963).

Proliferating normal endothelial cells have been shown to be more susceptible to infection with salivary gland virus (Vogel, 1958). However, viral infection of endothelial cells is not limited to either this particular virus or to proliferating endothelial cells since Coxsackie B_4 virus infection of these cells, leading to their necrosis and desquamation, has been identified in arteries and veins of mice (Sohal and Burch, 1969; Burch, Tsui and Harb, 1971). Mims (1968), also using mice, found that intravenously injected cowpox virus entered the endothelial cells of vessels considered to be mostly small venules and resulted in characteristic skin lesions. The specificity of this endothelial cell uptake was shown by the fact that no significant uptake by Küpffer cells in the liver occurred until massive doses of virus were given intravenously. Estes and Cheville (1970) identified a viral infection of endothelial cells in horses termed 'equine viral arteritis' which produced enormous hydropic swelling of the endothelial cells causing obliteration of the lumina of capillaries and anoxic necrosis of the smooth muscle of the small arteries. It also appears that in man many of the viruses of the exanthemata, such as varicella, vaccinia, variola, rubella and rubiola, and also certain of the arboviruses, can enter the blood stream and invade and damage endothelial cells (McKay and Margaretten, 1967). The resulting shedding of endothelial cells can then lead to intravascular thrombosis. The vascular lesions that occur as a result of human congenital rubella consist of intimal fibromuscular thickening (Esterley and Oppenheimer, 1967), but whether or not this is secondary to endothelial damage in the foetus is not possible to say.

AGING

The changes that occur in endothelial cells due to aging, or senescence, have not received as much attention as they rightfully deserve. Kuwabara and Cogan (1963) found that there was a progressive decline in the number of endothelial cells per unit length of retinal capillary vessel with advancing age. This is in keeping with the general scheme of senile atrophy of tissues in which is found a reduction in cell numbers with an increase in individual cell size (Tauchi, 1962). Kirk and Laursen (1955) found that the human aortic intima, in common with the other layers of its wall, increased its permeability to materials such as O_2, N_2, CO_2, glucose, lactate and iodine with age. Gerrity and Cliff (1972), in studying the morphological alterations associated with aging in the tunica intima of rats, found no increase in the numbers of endothelial cytoplasmic vesicles per unit area but did find increased surface area of the endothelial cells being produced by numerous folds and projections formed on the luminal surfaces of these cells. This increase in total area available for diffusion would tend to increase the net diffusion occurring across the membrane. In addition these workers found that in old rats the aortic endothelial cells developed large quantities of endoplasmic reticulum and Golgi substance. At the same time both osmiophilic and non-osmiophilic lipid inclusions appeared within the endothelial cells, whilst the endothelial nuclei became denser and extremely irregular in shape, with tortuous folded contours. Similar changes associated with aging have been identified in Purkinje cells (Andrew, 1962), hepatocytes (Tauchi, 1962) and aortic smooth muscle (Cliff, 1970). The finding of senile changes supports the conclusion of Spraragen, Bond and Dahl (1962) that the aortic endothelium is a very stable population of cells, since rapidly turning over cell populations do not show such senile changes (Andrew, 1962).

4

THE EXTRA-ENDOTHELIAL CELLS OF BLOOD VESSEL WALLS

The common feature shared by blood vessels is that their surfaces are lined by some type of endothelial membrane (see chapter 3). Apart from this structure their walls can vary considerably. In the simplest instances, as found at times in the small nutrient-net, or capillary vessels, the wall can consist solely of endothelial cells, with only basement membrane external to them (Landis, 1926; Nicoll and Frayser, 1967) (figure 16).

PERICYTES

Most vessels in these networks, however, have variable numbers of supporting cells present within their walls. It is likely that the number of names is greater than the number of cell types actually present. These cells have been variously identified as adventitial cells (Clark, 1936; Clark and Clark, 1939, 1940; Krogh, 1959; Cliff, 1963), peri-endothelial cells, or pericytes (Sandison, 1932; Fawcett, 1959, 1963; Majno and Palade, 1961; Farquhar and Palade, 1962; Cotran, Guttuta and Majno, 1965; Bruns and Palade, 1968a; Ludatscher and Stehbens, 1968; Rhodin, 1968; Crocker, Murad and Geer, 1970; Mikami et al., 1970; Grubb and Jones, 1971) or Rouget cells (Zweifach, 1937; Krogh, 1959; Hama, 1961). These cells share the features of being enclosed by a basement membrane system which at times is continuous with the endothelial basement membrane, where the two come into close enough relation, and of being longitudinally oriented, elongated cells with numerous tapering processes which encircle the vessel walls (Farquhar and Palade, 1962; Kuwabara and Cogan, 1963; Majno, 1965; Rhodin, 1968; Mikami et al., 1970) (figures 14, 27, 45). Both adventitial cells and pericytes are non-contractile cells which have well-developed phagocytic powers (Palade, 1960; Majno and Palade, 1961; Farquhar and Palade, 1962; Cliff, 1963; Cotran et al., 1965). It would appear that adventitial cells and pericytes (or peri-endothelial cells) are the same cell, being known by different names by different workers. The Rouget cell, however, can be distinguished from these by its contractile properties and its lack of phagocytic activity (Krogh, 1959). The validity of distinguishing these cells from vascular smooth muscle cells is, however, highly questionable, particularly in the face of evidence that vasomotion in vertebrates can only occur in vessels that have smooth muscle in their walls (Sandison, 1932; Clark and Clark, 1940, 1943; Chambers and Zweifach, 1944; Grafflin and Bagley, 1953; Nicoll and Webb, 1955; Lutz and Fulton, 1958). I am distinguishing here between vasomotion and the phenomenon of endothelial nuclear swelling that has been observed to occur in capillary vessels. This can alter the effective size of the capillary lumen by bulging the endothelial lining into it, as identified in both the liver (McCuskey, 1966; Burkel, 1970) and the rabbit ear chamber (Sanders, Ebert and Florey, 1940).

A beautiful description of the typical appearance of the numerous cytoplasmic extensions of pericytes that invest and support the underlying endothelial cells of the walls of minute blood vessels was given by K. W. Zimmerman in 1923, whose work has been brought to contemporary notice by Majno (1965) in his review of the ultrastructure of vascular membranes. Zimmerman used the technique of silver impregnation to demonstrate the morphology of these cells with considerable precision, when judged by present day electron microscopic descriptions. However, the specificity of his technique was not sufficient to differentiate between true pericytes and the epithelial cells of the renal glomerulus (see Majno, 1965), which are clearly distinguishable by electron microscopy (Farquhar and Palade, 1962).

The fine structure of pericytes is somewhat variable and generally rather undistinguished. An elongated nucleus shows the same general curvature about the vessel wall as the cell as a whole and characteristically forms a bulge on the abluminal surface of the cell. Very commonly, sections through the branching processes and cell bodies of pericytes contain no nuclear profiles. Nucleolar material is generally not identified. Small numbers, of oval mitochondrial profiles and short segments of both rough and smooth endoplasmic reticulum are present in the perikaryon at the nuclear poles. Golgi apparatus is scant. The cytoplasm usually contains small quantities of fine fibrils and microtubules. The plasma membrane has moderate numbers of caveolae, or surface-connected vesicles present. Small numbers of membrane-bounded dense bodies may be present within the cytoplasm. In general the cells have an undifferentiated appearance (Rhodin, 1968) and as such could possibly correspond to the 'mesenchymal cells' of Maximow (Arey, 1936) which are considered to reside in such situations and to be activated during inflammation and wound healing (Hadfield, 1951; Crocker *et al.*, 1970). There is no doubt that these cells are capable of mitotic division and that they can modulate into vascular smooth muscle cells in the course of blood vessel differentiation during wound healing (Clark and Clark, 1940; Cliff, 1963, 1965).

The similarity between endothelial cells and pericytes is often very striking, and Fawcett (1959) in fact considered that they were identical in their electron microscopic appearances. Indeed the similarity seen in morphological terms between these two cell types can also, at times extend to identical function. Schoefl (1963), studying proliferating blood capillaries electron microscopically, found that the same cell in one part of the section was part of the endothelium lining a newly formed blood vessel whilst in another region of the section it was quite definitely a pericyte. Her observations are supported by those of Farquhar and Palade (1962) who observed that pericytes in the renal glomerulus extended bulbous processes into the vascular lumen via narrow necks penetrating between the endothelial cells to make contact with the blood stream. The function of these processes is obscure. Pericytes form cell-to-cell attachments of the zonula adhaerens type both between themselves (Farquhar and Palade, 1962) and with endothelial cells in regions where the basement membranes are absent (Rhodin, 1968; Crocker et al., 1970). This latter form of cell-to-cell relation is considered to be of importance in providing a form of contact inhibition whereby pericytes that become incorporated into the walls of newly formed vessels can suppress the formation of further endothelial outgrowth. This interesting concept finds support in the conclusions reached by Kuwabara and Cogan (1963) concerning the inhibitory effect of mural cells on new vessel formation in the retina. These authors found evidence for mural cells forming contacts with endothelial cells but went to some length to differentiate the mural cells of retinal capillaries from pericytes. I do not consider that these cells should be distinguished from the general class of pericytes (a view also taken by Majno, 1965), particularly as their presence within the vascular basement membrane system was Kuwabara and Cogan's major reason for excluding them from this cell class.

Embryologically, pericytes are formed by the incorporation of cells from the surrounding mesenchyme into the walls of primitive blood vessels (Sethi and Brookes, 1971), whilst in new vessel formation, occurring in inflammation and the healing reaction, such cells appear to be derived mainly from proliferating fibroblasts incorporated into the walls of the newly formed vessels (Clark and Clark, 1939, 1940; Cliff, 1963). In addition, pericytes are also produced during wound healing by mitotic division of existing pericytes (Clark and Clark, 1940). Crocker and others (1970) claim to have found evidence for the existence of primitive mesenchymal cells that become incorporated into the basement membrane systems of the newly formed vessels of healing tissue to become pericytes.

Considerable attention has been given to the in-vivo properties of the pericyte in studies on the microcirculation, mainly in amphibians and mammals. It is sometimes considered that although capillary contractility does not exist in mammals it may occur in amphibians (Clark and Clark, 1943), although even in these animals there is considerable doubt

(Chambers and Zweifach, 1944; Grafflin and Bagley, 1953; Lutz and Fulton, 1958). The possible role of the peri-endothelial 'Rouget cells' in producing such capillary contractions in amphibians appears to be even more discountenanced by the observations of Hama (1961) who found myofibril-type structures within the endothelial cells themselves but not within the Rouget cells of amphibians. Zweifach (1937) made an appraisal of the relations of Rouget cells to capillary contractility in a paper dealing with the in-vivo properties of amphibian capillaries, which also took into account a good deal of the early literature on the subject, and concluded that these cells were not contractile.

The histochemistry of pericytes has been investigated to some degree, for instance the demonstration of lactate–DPN and isocitrate–TPN dehydrogenase activity by Kuwabara and Cogan (1963) enabled them to stain the cell bodies and processes of pericytes with considerable specificity. By contrast, the endothelial cells reacted strongly for alkaline phosphatase activity whilst the pericytes were negative for this reaction.

It would appear that pericytes have primarily a mechanical supporting function within the walls of minute blood vessels, performed by their investment of the endothelium with numerous elongated processes and also by their production of additional basement membrane condensations external to the endothelium. Another role performed by pericytes is that of phagocytosis of material within the vessel wall; however, in this regard these cells are generally overshadowed by the perivascular macrophages (Cotran *et al.*, 1965) (see p. 93). A final possibility is that these cells may represent a source of undifferentiated mesenchymal cells which can participate in repair processes and inflammation.

These basic properties of pericytes are varied in certain specialized situations. In the mammalian retina the pericytes, termed 'mural cells' by Kuwabara and Cogan, are considered to be distinguished by the presence of long finger-like processes that extend longitudinally or else spirally around the vessel wall and their function is thought to be in the regulation and distribution of blood flow through the very rich capillary networks of the retina. The loss of mural cells from the walls of the retinal capillaries that occurs during diabetes mellitus can be associated with the loss of vessel tone, establishment of dilated thoroughfare channels, microaneurysms, and all the other features of diabetic retinopathy (Kuwabara and Cogan, 1963).

The eyes of certain fish possess a vascular structure termed the choroid *rete* which is considered to be involved in the maintenance of the very high partial pressures of oxygen in the vitreous humour related to their retinae (Wittenberg and Wittenberg, 1961). Fawcett (1963) discussed the structure and function of this vascular rete and pointed out that the plump afferent vessels of the rete have prominent pericytes, whilst the thin-walled efferent vessels with which they are intimately related have no pericytes.

The counter-current system of this rete apparently has both an oxygen and a water exchange role. A similar pattern of distribution of pericytes, being present in the afferent and absent in the efferent vessels, is also found in the vascular rete mirabile of the gas gland in the swim bladder of certain teleosts (Fawcett, 1963). These rete apparently have very similar functions to the choroid rete (Wittenberg and Wittenberg, 1961).

The pericytes in the mammalian kidney show interesting morphological and functional differentiation both in the glomeruli (Farquhar and Palade, 1962) and in relation to the vasa recta of the outer medulla, which once again are involved in counter-current exchange processes analogous to those of the rete mirabile (Berliner et al., 1958). Each medullary pericyte is related to both ascending and descending vasa recta and shows striking specialization, as described by Moffat (1967). The pericyte cytoplasm, where it is related to the plump descending vasa recta, contains tracts of intracytoplasmic fibrils, termed by him 'myofibrils', whilst on the opposite side of the pericyte, where it is related to the thin fenestrated endothelium of the ascending vasa recta, the cytoplasm has no fibrils but has numerous pinocytic vesicles on the plasma membrane. Moffat has suggested that the pericytes control medullary blood flow through the smooth muscle-like regions which are related to the descending vascular limbs and that they react both to humoral factors and to local sampling by the pinocytic vesicles which are related to the ascending vascular limbs.

Hence, to the functions of pericytes summarized above, may be added their participation in various counter-current vascular exchange systems by means which, at the moment, are obscure.

VASCULAR SMOOTH MUSCLE CELLS

Smooth muscle cells are very conspicuous and important components of blood vessel walls, where they form the major and in many cases the sole cellular element of the tunica media. In general, they are rather large elongated cells being about 2.0 μm in width and 60 μm in length (Rhodin, 1962a). In muscular arteries and arterioles they are spindle shaped with long tapering ends. In elastic arteries they have highly irregular shapes with numerous branched processes forming contacts with both the elastic elements of the media and related cells (Cliff, 1967, 1970) (figure 28). They have been identified as the sole cellular component in the media of the elastic arteries of mammals (Keech, 1960a; Pease and Paule, 1960; Wolinsky and Glagov, 1964; Chan, Balis and Conen, 1965; Haust et al., 1965; Cliff, 1967, 1970; Stein, Eisenberg and Stein, 1969). In birds similar cells form the major medial population in these vessels (Karrer, 1960b; Moss and Benditt, 1970) but a second type of non-muscular, generally rather nondescript cell has been identified within the fibro-elastic

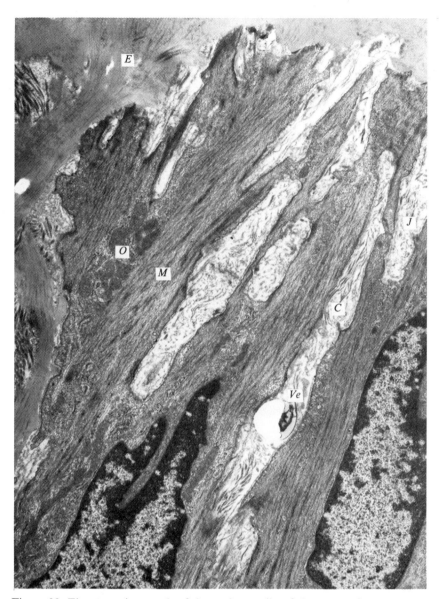

Figure 28. Electron micrograph of the tunica media of the aorta of a mature rat containing smooth muscle cells whose cytoplasm is filled mainly with obliquely oriented tracts of myofibrils (*M*) that terminate at points of attachment of the cells with elastic tissue (*E*). These myofibril tracts are separated by organelle-rich regions of cytoplasm (*O*). The cells have numerous processes that make contact with the elastica and also with one another (*J*). *C*, collagen; *Ve*, caveolae on plasma membrane. Magnification × 10000.

74 The extra-endothelial cells of blood vessel walls

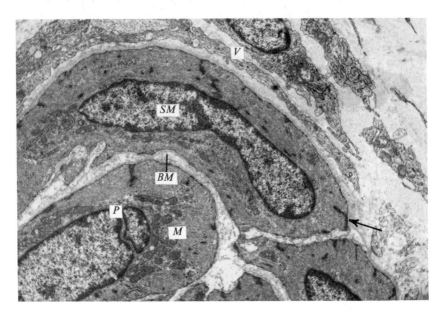

Figure 29. Electron micrograph of a small muscular artery in a rabbit containing smooth muscle cells (*SM*) in its media. Their cytoplasm is almost completely filled with myofibrils which have regions of increased electron density in relation to the plasma membrane (arrow). Spindle-shaped dense regions occur within the myofibril tracts as they run through the cytoplasm. Membranous organelles, mainly mitochondria (*M*) are concentrated in the perikaryon. A veil cell (*V*) lies external to the smooth muscle of the media. *P*, nuclear pinch; *BM*, smooth muscle basement membrane. Magnification × 7500.

interlamellar spaces (Moss and Benditt, 1970). Circularly or spirally oriented smooth muscle cells form the major component of the media of the muscular arteries of all sizes (Moore and Ruska, 1957; Pease, Molinari and Kershaw, 1958; Pease and Molinari, 1960; Zelander, Ekholm and Edlund, 1962; Hogan and Feeney, 1963) (figure 29) and of arterioles (Nicoll and Webb, 1955; Ancla and de Brux, 1964; Van Citters, 1966; Rhodin, 1967; Burkel, 1970) (figures 5(*a*), 30) where they extend into the walls of the smallest such vessels. The final, extremely narrow precapillary ramifications of the arterial system, which arise at angles of about 90° from the terminal arterioles, have no smooth muscle cells in their walls and their flow is controlled by sphincter-like collections of smooth muscle cells present at their points of origin from the arterioles (Chambers and Zweifach, 1944; Rhodin, 1967; Burkel, 1970). These vessels I would term 'metarterioles' by virtue of their arteriolar form, course and calibre but lack of smooth muscle in their walls (figure 5(*a*)). Chambers and Zweifach (1944), however, used this term for the terminal arterioles which still have definite smooth muscle present in their

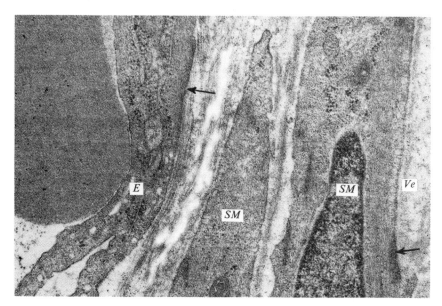

Figure 30. Electron micrograph of an arteriole in a rabbit. It has a media composed of two layers of smooth muscle cells (*SM*) whose cytoplasm is mainly occupied by myofibrils with spindle-shaped dense regions along their lengths. Both these fibrils and similar fibrils in the endothelial cell (*E*) terminate in electron-dense regions at the plasma membranes (arrows). *Ve*, caveolae on the plasma membrane. Magnification × 30 000.

walls and termed the sphincters that control the ostia of the branches arising from their sides 'precapillary sphincters'.

The metarterioles, beyond the sphincters controlling their ostia, and the capillary networks have no smooth muscle in their walls, the medial coat here being represented by the pericytes (see p. 68). Medial smooth muscle cells begin to appear in the venous vessels of the bat's wing at the level of the smallest veins at the site of the first valves (Nicoll and Webb, 1946). Clark and Clark (1943) noted neuromuscular control of the calibre of minute veins during observations on blood vessels in the rabbit ear chamber. Rhodin (1968) performed a most useful systematic study of the fine structure of the minute vessels identifiable in the fascia of the medial thigh muscles of rabbits. In the media of collecting venules which had luminal diameters of 30–50 μm he was able to identify some cells as primitive smooth muscle cells with poorly developed myofibril tracts. The muscular venules, with luminal diameters of 50–100 μm and walls approximately 2.0 μm in thickness, had muscular medial coats from one cell to two cells thick. The smooth muscle cells for the most part were arranged spirally with a very coarse pitch, being almost longitudinal in orientation in these vessels.

The smooth muscle cells within the media of veins show considerably greater variation in their distribution and disposition than those of the arterial vessels. Hence the large retinal veins of the human being, the monkey and the rat have no smooth muscle in their walls, whilst the corresponding retinal arteries do (Hogan and Feeney, 1963). Similarly, venous limbs of arteriovenous anastomoses, as studied in human nasal mucosa by Cauna (1970), are completely devoid of muscular elements. Franklin (1928), in a very comprehensive review of the structure and function of veins, pointed out that veins in muscle, bone and deep organs in general have little intrinsic smooth muscle and that deep veins running in relation to the arterial vessels between muscle masses have moderate amounts of intrinsic smooth muscle, whilst superficial veins have the greatest development of intrinsic smooth muscle in their walls. The effect of gravity was reflected in the greater development of the smooth muscle coats in the veins of the legs as compared to those in the arms. He stressed the great distensibility of the veins of various diving animals and instanced the case of the Great Northern Diver (a bird) in which the venae cavae and venae cavae hepaticae were the same size, when dilated, as the corresponding vessels in the human being! He assigned to the circular muscle fibres of such veins a constrictor action and to the longitudinal muscle fibres a dilator action.

Quite distinct inner circular and outer longitudinal muscle coats are present in the hepatic portal vein of the rat (Voth *et al.*, 1969; Ts'ao, Glagov and Kelsey, 1970) and the anterior mesenteric veins of rabbits, cats and guinea pigs and cattle (Sutter, 1965; McConnell and Roddie, 1970). It is interesting to note that the spontaneous rhythmic contraction that occurs in these vessels (figure 7) appears to be largely a feature of the external longitudinal muscle coat. Such spontaneous activity of veins is remarkably widespread with regard to both species and vessels involved (see Franklin, 1928 for review). Perhaps the best known and certainly one of the most dramatic demonstrations of such venous activity is seen *in vivo* in the bat's wing membrane (Nicoll and Webb, 1955). These veins also have distinct circular and longitudinal coats of smooth muscle (Nicoll and Webb, 1946).

As a final example of the great variability of the intrinsic musculature of vein walls I would like to cite the work of Alexander and Jensen (1963) who found that the pulmonary veins of cattle had very heavy muscular media right down to vessels of 20 μm diameter. In veins of less than 300–400 μm diameter this heavy medial coat was characterized by abrupt discontinuities, producing 'sphincter-like' discrete smooth muscle masses in the wall, which had the appearance of being spirally arranged. Casts made of the veins confirmed the presence of annular projections of the walls extending into their lumina. What this adaption in the pulmonary veins of cattle achieves is unknown, but it certainly warrants further investigation.

Smooth muscle cells are also found in certain instances as normal constituents of the tunica intima of arteries. The longitudinally oriented smooth muscle of the intima of the arterial limbs of arteriovenous anastomoses (Cauna, 1970), of the intimal cushions that occur distal to branches of cerebral arteries in children (Hassler, 1962), and of the spirally arranged intimal cushions of human umbilical arteries (Monie, 1945) all appear to be physiological, as opposed to pathological, developments. Arteriovenous anastomoses are characterized by an 'all-or-none' type of flow pattern (Burton, 1954) and as the action of circular smooth muscle alone within a vessel wall cannot effect complete closure of the lumen in any but the smallest arterioles (Van Citters, 1966) it becomes necessary to have longitudinally acting muscle elements present to increase the wall thickness by acting to shorten it, as expounded by Roach (1970). A physiological function of these structures in the cerebral circulation has been suggested by their rich innervation (Hassler, 1962). The spiral intimal cushions of umbilical arteries are considered to be of rheological importance in adapting this vessel to the great variations in blood flow that occur therein, most particularly during labour. It is quite feasible that the highly developed intimal smooth muscle layers, both longitudinal and circular, present within human coronary arteries (see French, 1970) may be of similar rheological significance in such vessels which show extreme fluctuations in rate and even in direction of blood flow (Gregg, 1934).

The large arteries of many mammals contain intimal smooth muscle cells which have a longitudinal orientation (French, 1970). In the rabbit aorta such cells have been equated with the Langhans cells of the human aortic intima (Seifert, 1963) and it is presumed that it is these same cells that were described by Lautsch, McMillan and Duff (1953), in whole mounts of human aortic intima, as large, branching anastomosing cells. Lee and others (1970) found smooth muscle cells normally populating the aortic intima of swine while Gerrity and Cliff (1972) found smooth muscle cells being incorporated into the aortic intima of rats during normal growth processes. A similar 'liberation' of smooth muscle cells into the intima of arteries by the breakdown of the internal elastic lamina has been described occurring in rabbits (Cookson, 1971).

In general, the fine structure of vascular smooth muscle cells is similar to that of smooth muscle in other sites. Whilst Rhodin (1962a) was able to discern certain differences between vascular and intestinal smooth muscle cells, he considered that there were more features of similarity than dissimilarity between the various forms of smooth muscle cells. Each cell has a large centrally located nucleus with peripheral chromatin clumping and generally moderately developed nucleolar material. The outline of the nucleus may be extremely irregular, with numerous folds and pinches present (figure 29), this being taken as morphological evidence of contraction (Policard, Collet and Giltaire-Ralyte, 1955; Buck,

1961; Rhodin, 1962a; Cliff, 1967). The nuclear length-to-width ratio may change from 6–8:1 in relaxed to 1:1 in contracted smooth muscle (Somlyo and Somlyo, 1968). Folds and pinches in the nuclear outline are considered to be produced, at least in part, by the action of contractile filaments attached to the outer layer of the nuclear membrane (Franke, 1970). These nuclear stigmata of cellular contraction are quite distinct from the age-associated development of fissured multilobed nuclei with a 'tortured' appearance, identified in the aortic media of senile rats (Cliff, 1970). The nucleus is contained by a double unit membrane envelope which encloses the perinuclear canal. The inner and outer membranes of the nuclear envelope are continuous at the margins of nuclear pores.

The cytoplasm of vascular smooth muscle cells contains all the usual membranous organelles which, in the case of the spindle-shaped cells, are localized mainly to the thickest regions of the cell body at the two poles of the nucleus (Pease and Molinari, 1960; Rhodin, 1962a) and, additionally, in the serrated, irregularly branched cells of the media of elastic arteries, to the organelle-rich zones that lie between the oblique myofibril tracts that are characteristic of these cells (Cliff, 1967, 1970) (figures 28, 31).

Mitochondrial profiles in vascular smooth muscle cells are fairly numerous (Policard et al., 1955; Pease and Molinari, 1960) and are generally oval to moderately elongated in shape, with their long axes aligned with the general orientation of the fibrillar cytoplasm (Cliff, 1967, 1970). Elements of smooth endoplasmic reticulum are not very conspicuous in vascular smooth muscle (Rhodin, 1962a). It has been suggested that apparent connections between elements of the endoplasmic reticulum and caveolae on the plasma membrane may have a role in the movement of ions within smooth muscle cells (Burnstock and Merrillees, 1964) which in effect could be analogous in these cells to the 'T' system of striated muscle in the storage and release of calcium ions (Gabella, 1971). Ergastoplasmic elements (or 'rough' ER) are considerably more common in vascular smooth muscle than in most other forms of smooth muscle (Rhodin, 1962a). This cytoplasmic component is very highly developed in the cells of the tunica media of elastic arteries of developing animals (Simpson and Harms, 1964; Chan, Balis and Conen, 1965; Balis, Chan and Conen, 1967; Cliff, 1967), so plentiful, in fact, that at times it has led to the identification of these cells as fibroblasts (Karrer, 1960b). However, as development proceeds and the media attains its mature form the cells progressively lose their 'fibroblastic' ergastoplasm and become more and more 'muscular' in appearance (Karrer, 1960b; Chan et al., 1965; Cliff, 1967; Stein et al., 1969) (figure 28).

In senile animals the smooth muscle cells of the aortic media once again develop considerable quantities of ergastoplasm (Cliff, 1970).

A Golgi system is present in the perikaryon of vascular smooth muscle cells (Rhodin, 1962a) but, as is the case with the ergastoplasm, it is well-

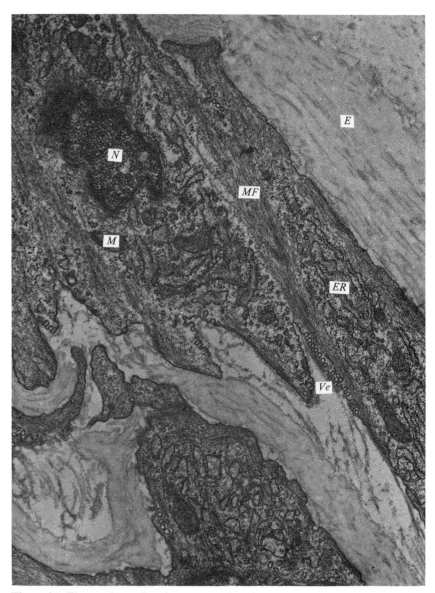

Figure 31. The aortic media of a one-month-old rat. The smooth muscle cell cytoplasm has very characteristic alternating oblique tracts that are occupied either by myofibrils (*MF*) or by membranous organelles, mainly ergastoplasm (*ER*) and mitochondria (*M*). *E*, elastica; *N*, nucleus; *Ve*, caveolae on plasma membrane. Magnification ×24000. Micrograph supplied through the kindness of Dr R. G. Gerrity.

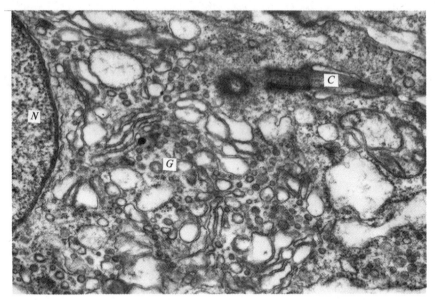

Figure 32. A smooth muscle cell in the aortic media of a 2-week-old rat has a cilium (*C*) extending from its surface. Paired centrioles form a typical basal body. A large Golgi region (*G*) contains smooth-surfaced membranes, in paired stacks, and numerous vesicles. *N*, nucleus. Magnification × 30000. (Cliff, 1967.)

developed only in the media of elastic arteries of growing (Cliff, 1967) and of senile animals (Cliff, 1970) (figure 32). Paired centrioles can be identified in relation to the Golgi system and also, occasionally, in young animals the smooth muscle of the aortic tunica media may have paired centrioles forming the basal body of a cilium extending from the cell surface (Cliff, 1967; Gerrity, 1972) (figure 32). Cilia have been identified in other forms of smooth muscle cells and fibroblasts by Sorokin (1962).

Other membranous organelles, including multivesicular bodies, smooth and coated vesicles and probable lysosomes containing lipid and granular material, can also be identified in vascular smooth muscle cells (Cliff, 1970). Free ribosomes are present in the cytoplasm of smooth muscle cells of developing vessels, usually in chains or clusters related to developing myofibril tracts (Chan et al., 1965; Balis et al., 1967). Zelander and others (1962) have described concentrations of granular dense materials related to indented regions of the plasma membrane in smooth muscle cells of pancreatic arteries.

The major cytoplasmic components of mature vascular smooth muscle cells are the myofibril tracts. These tracts contain finely fibrillar material which is aligned to quite a high degree with their long axes. The tracts terminate at each end in zones of increased electron-density related to the plasma membrane, and scattered throughout their length are interspersed spindle-shaped zones of increased electron density termed 'dense bodies' (Pease and Molinari, 1960; Rhodin, 1962a) (figure 29). Both these features are highly characteristic of smooth muscle cells in general. The majority of fibrils within the cytoplasmic tracts of vascular smooth muscle are 7.0–8.0 nm in diameter (Rhodin, 1962a; Cliff, 1967). Although other workers have found lower values of 3.0–4.0 nm (Pease and Molinari, 1960; Karrer, 1961), Rhodin did not consider there was proof of the existence of two sets of filaments, as found in striated muscle, even though he could identify small numbers of 5.0 nm filaments in addition to 8.0 nm filaments (above). This failure to demonstrate a double population of filaments is not peculiar to vascular smooth muscle but is a feature of all smooth muscle cells (Somlyo and Somlyo, 1968). Lowy and Hanson (1962) have suggested the possibility that in vertebrate smooth muscle actin and myosin fibrils may both have the same diameters; Rüegg (1971), in considering the elusive thick myosin filaments, suggested that myosin was probably present in a soluble dispersed form within smooth muscle cells. Immunofluorescent 'staining' of both human and bovine blood vessel walls with specific antimyosin and anti-actomyosin leaves no doubt as to the presence of myosin and actomyosin within the vascular smooth muscle cells (Knieriem, Kao and Wissler, 1967, 1968; Becker and Murphy, 1969). However, since the sliding filament hypothesis as proposed for the contraction of striated muscle (Huxley, 1957) is also considered to be operative in smooth muscle cells (Somlyo and Somlyo, 1968; Lowy and Small, 1970) great efforts have been made to demonstrate 'thick' myosin filaments in addition to the 'thin' (5.0–8.0 nm) actin filaments. Stretching the smooth muscle to about 150% of resting length appears to be an important factor in demonstrating the thick filaments. Garamvölgyi, Vizi and Knoll (1971) employed this technique to demonstrate 13 nm filaments in sheets of smooth muscle cells dissected from the ileum of the guinea pig. Stretching and incubation for some time before fixation was used to demonstrate myosin 'ribbons' and 5.0 nm actin filaments in *taenia coli* (Lowy and Small, 1970) and also thick 18 nm and thin 8.4 nm diameter filaments in vascular smooth muscle of the portal vein (Devine and Somlyo, 1971). Glycerination before fixation has also been successful in demonstrating two sets of fibrils. Kelly and Rice (1968) used extremely prolonged glycerine soaking, up to six months, before fixation of gizzard smooth muscle to demonstrate both 11–21 nm and 5.0–7.0 nm diameter sets of fibrils. They emphasized that at a pH above 6.6 myosin would be solubilized (see also Somlyo and Somlyo, 1968). Keyserlingk (1970) used glycerination together with ATP-induced contraction to produce a population of thick (15–20 nm diameter) filaments in gut smooth muscle. As an alternative to these attempts to demonstrate a double population of filaments another form of association between actin and myosin has been suggested by Panner and Honig (1970) who claimed to resolve separate individual myosin molecules with characteristic heads and tails in ultrathin sections associated with 6.0 nm diameter actin filaments.

In addition to the myofilaments discussed above there has been demonstrated within smooth muscle cells a separate discrete population of 10 nm filaments within smooth muscle cells which are considerably more resistant to extraction and more readily fixed than the myofilaments (Uehara, Campbell and Burnstock, 1971; Cooke and Fay, 1972). These filaments appear to be linked through the fusiform dense bodies into a three-dimensional framework within the cell and to be anchored at each pole of the cell. On stretching, this framework is drawn into the central region of the cell and presumably it functions to resist overextension. The fusiform dense bodies are apparently composed of protein and have been demonstrated by differential extraction to be quite distinct from the myofibrils with which they are so intimately associated (Rhodin, 1962a; Keyserlingk, 1970).

The plasma membrane of vascular smooth muscle cells is of the usual trilaminar unit membrane type (Robertson, 1959). Numerous *caveolae intracellulares* or pinocytic vesicles, 65 nm in diameter, are present on the plasma membranes (Pease and Molinari, 1960; Rhodin, 1962a; Cliff, 1967, 1970), but are restricted in distribution to those regions of the membranes that are unrelated to the terminations of the myofibril tracts (figure 33). Each vascular smooth muscle cell is invested more or less completely by a basement membrane system similar to that which invests other forms of smooth muscle cells (Caesar, Edwards and Ruska, 1957) (figure 29). The basement membranes of adjacent smooth muscle cells frequently merge and appear to have a function in binding together the cells of the media (Policard et al., 1955; Pease and Molinari, 1960). In certain instances the smooth muscle basement membranes fuse with the endothelial basement membranes (Pease et al., 1958) whilst in other cases these two systems remain quite distinct (Zelander et al., 1962). The muscle cells of retinal arteries, which are devoid of elastic tissue, have extremely well-developed basement membranes, which at times show a laminated structure (Hogan and Feeney, 1963). A close topographical association between the basement membranes of the media and developing elastic tissue has frequently been noted (Pease et al., 1958; Pease and Molinari, 1960; Paule, 1963; Haust et al., 1965; Cliff, 1967). In addition, in elastic arteries the muscle cell basement membranes frequently merge with the related elastic tissue fibres and in such regions can no longer be indentified. In such regions smooth muscle cells can form attachments, characterized by 15–20 nm separation, with elements of the elastic tissue present in both the laminae and the interlaminar tracts (Keech, 1960a; Pease and Paule, 1960; Cliff, 1967). The smooth muscle cells in such vessels appear to act as 'span cells' linking elements of

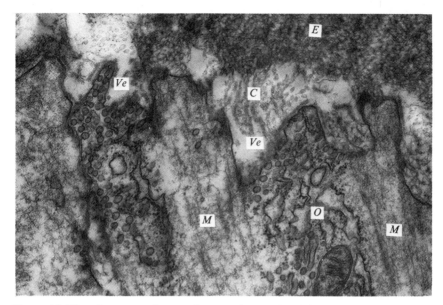

Figure 33. The plasma membrane of a smooth muscle cell in the media of the rat's aorta has regions where numerous surface-connected vesicles (*Ve*) are directly related to the organelle-rich cytoplasmic tracts (*O*). By contrast the plasma membrane related to the myofibril tracts (*M*) have no caveolae. The elastica (*E*) can be seen to have a fibrillar network structure. *C*, collagen. Magnification × 30 000.

the fibro-elastic stroma within the wall (Burton, 1954). Regions of attachment to elastic elements on opposite sides of the smooth muscle cells are linked by relatively short obliquely oriented myofibril tracts within their cytoplasm (figure 31). This results in a 'myo-elastic' form of tissue organization (Hass, 1939a, b).

As well as such attachments to elastic tissue, the muscle cells of blood vessel walls also form attachments both with one another and with the endothelium. These contacts occur in regions where the basement membrane systems are deficient and may take one of several forms. In the simplest type there is found a region of parallelism between the related plasma membranes which are separated by a gap of 20 nm. This occurs in muscle-to-muscle (Pease and Paule, 1960; Cliff, 1967; Rhodin, 1967) and endothelium-to-muscle attachments (Parker, 1958; Rhodin, 1967, 1968; Gerrity and Cliff, 1972). In this latter form of contact the endothelial cell usually sends a process down through an opening in the internal elastic lamina to make contact with the muscle in the media (figure 21). Such contacts are considered to have significance in supplying nutriments directly to the inner media via the endothelium (Parker, 1958). 'Peg and socket' forms of interrelationship are also identified between the muscle cells of avian blood vessel walls (Bolton and Nishihara, 1970). It was

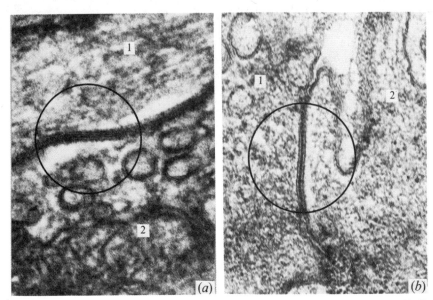

Figure 34. Two examples of nexus regions (circled) between smooth muscle cells (1 and 2) of the rat's aortic media. (*a*) A region where the nexus is 7-layered, magnification × 170000. (*b*) A region where the nexus appears to be 5-layered, magnification × 120000.

considered that the low resistance pathways for electrical currents detected were related to the presence of these structures and to frequent regions of membrane nexus present between the vascular smooth muscle cells. Membrane nexus regions, which are characterized by apparent fusion of the outer leaflets of the trilaminar plasma membranes of related cells to produce quintuple-layered membranes (Dewey and Barr, 1962; Lane and Rhodin, 1964; Oosaki and Ishii, 1964), have also been identified in the media of mammalian vessels (Cliff, 1967, 1970) (figure 34). The excellent study of nexus regions in ureteric smooth muscle by Uehara and Burnstock (1970) has shown that in fact 2.5–3.0 nm gaps exist between the related plasma membranes so that, in fact, seven layers can be counted in such forms of membrane adhesion. Nexus regions are considered to act as sites for electronic coupling of smooth muscle cells (Dewey and Barr, 1962), which, as in the case of those in blood vessel walls, are of the single-unit or 'visceral' type in terms of their electrophysiology (Lane and Rhodin, 1964; Burnstock, Gannon and Iwayama, 1970; Ljung, 1970).

Vascular smooth muscle cells are multifunctional cells (Dunihue, 1967; Wissler, 1967). The contractile activity of these cells is best seen in the active changes that occur in vessel diameters associated with vasomotion, of which examples have been cited in preceding sections. Some measure of their ability to contract can be appreciated from observations on spiral

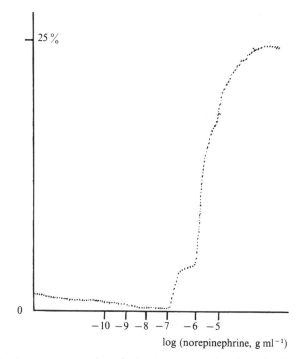

Figure 35. Kymograph tracing obtained from a strip of renal artery from a sixty-year-old woman. The abscissa shows the log dose of norepinephrine and the ordinate shows shortening as a percentage of original strip length. Time interval between doses were approximately 2 minutes.

strips cut from vessel walls which in optimal conditions can show up to 30 % reduction in length (Furchgott and Bhadrakom, 1958; Somlyo and Somlyo, 1964) (figure 35). Bohr, Filo and Guthe (1962) studied ATPase activity of glycerinated preparations of vascular smooth muscle from a range of vessels and concluded that vascular preparations displayed many of the features of other glycerine-extracted muscle models. Somlyo and Somlyo (1968), in considering the rather low content of actomyosin of vascular smooth muscle (2–10 mg per g wet wt) as compared to that of skeletal muscle (70 mg per g wet wt), suggested that this indicated a greater efficiency of the contractile mechanism of the former, in terms of tensions generated on a weight for weight basis.

The smooth muscle cells of blood vessels may respond to a variety of control mechanisms. These include adrenergic and cholinergic nerve fibres (see section on innervation in chapter 6), circulating vaso-active materials, mechanical stimulation and local chemical changes in the surrounding tissues. In-vivo responses of vascular smooth muscle to these various stimuli have been studied both in the microcirculation (Nicoll and Webb, 1955; Lutz and Fulton, 1958; Altura and Zweifach, 1965; Altura,

Figure 36. Kymograph tracing from a strip of aorta of a one-month-old rat. There is a slow rhythmical (7 per min) contraction of slightly more than 1 % of the resting strip length in oxygenated Krebs–Ringer solution.

1970) and in the arteries (Aars, 1971; Arndt et al., 1971) and veins (Guntheroth and Chakmakjian, 1971). An excellent review of the general control of vascular smooth muscle has been written by Mellander and Johansson (1968). Vane (1969) has provided a very comprehensive review of the action of a large number of circulating vaso-active substances affecting these cells, which includes acetylcholine, bradykinin, the catecholamines, histamine, serotonin, angiotensin, vasopressin and the prostaglandins. Recently, Burnstock (1972) has provided evidence for the existence of purinergic endings as a component of the autonomic nervous system. Direct vasodilator activity of adenosine compounds has long been recognized (Hilton, 1962), and Burnstock has suggested that purinergic fibres may have a vasodilator role.

The rhythmical phasic contractions of various veins has been described in previous sections but mention must be made of similar phasic activity that has been observed in tissue bath experiments with arterial strips from subcutaneous arteries of the dog (Johansson and Bohr, 1966), umbilical artery (Somlyo, Woo and Somlyo, 1965) and the rat's aorta (figure 36). Such spontaneous activity is probably of significance in arterial auto-regulatory mechanisms and underlines the functional connection of the cells of the media into a 'single-unit' type of smooth muscle for which morphological evidence has been given above. The possibility that cilia on the smooth muscle cells may have some form of mechanoreceptor role is worth considering (Sorokin, 1962).

The involvement of smooth muscle cells in fibroblastic activities in the walls of blood vessels has been suggested or inferred by numerous writers. These cells have been considered to be the sites of synthesis of both the collagen and the elastic fibres that are laid down within the media of developing arteries (Fawcett, 1959; Pease and Molinari, 1960; Cliff, 1967). The rough endoplasmic reticulum and Golgi apparatus present within such cells (p. 78) are features of protein synthesizing and secreting cells (Zeigel and Dalton, 1962; Wissig, 1963; Caro and Palade, 1964). Recently, this fibroblastic activity has been investigated by Ross and Klebanoff (1971) using [^{3}H]proline as a label for autoradiography of both oestrogen-stimulated uterus and the aorta in the developing rat. The smooth muscle cells in both sites showed rapid accumulation of label which subsequently shifted after an interval of 4 hr to be located over the connective tissue elements. Gerrity (1972) has extended such autoradiographic experiments considerably by combining this technique with morphometric stereology (Weibel, Kistler and Scherle, 1966; Weibel and Elias, 1967) and biochemical extraction and estimation of [^{3}H]proline and [^{3}H]hydroxyproline levels in the aorta of the growing rat. He has been able to demonstrate that the smooth muscle cells of the media concentrate [^{3}H]proline, and that [^{3}H]proline is converted to [^{3}H]hydroxyproline, which is only found in significant amounts in collagen and in smaller amounts in elastin (Stetten, 1949; Udenfriend, 1966; Bentley and Hanson, 1969) and in no other animal proteins. He has shown that the endoplasmic reticulum and Golgi apparatus are heavily labelled within the medial muscle cells and that the label subsequently moves out from the cells into recognizable collagen and elastic fibres (figure 37).

It is possible to culture vascular smooth muscle cells *in vitro* and to distinguish such cells quite clearly from connective tissue fibroblasts when these are also present in the explant (Kasai and Pollak, 1964). In culture the smooth muscle of the aortic media of the pig and the rabbit lose a large proportion of their myofilaments and *pari passu* develop large amounts of ergastoplasm to become fibroblastic in appearance (Fritz, Jarmolych and Daoud, 1970). Bloom (1930) cultured foetal guinea pig heart and aorta and demonstrated that both collagen and elastic fibres were produced in the cultures. The elastic tissue formed most prominently along lines of pull produced by the beating cardiac explants but were also found in cultures of aortic tissue alone. Recently, Ross (1971) found 'elastin-like' material in relation to cells cultured from the media and intima of aortae from prepubertal guinea pigs.

In mature animals vascular smooth muscle cells largely lose their fibroblastic regions and their cytoplasm is occupied almost exclusively by myofibril tracts and associated mitochondria (Parker, 1958; Pease and Molinari, 1960; Cliff, 1967, 1970; Stein *et al.*, 1969). However, they can be stimulated during repair processes to revert to their earlier type of structure, with well-developed ergastoplasm and Golgi systems present in their cytoplasm, in the media (Murray, Schrodt and Berg, 1966) and intima (Still and Dennison, 1967; Poole, Cromwell and Benditt, 1970) of arteries. Similar changes can be found in smooth muscle cells present in the arterial intima in spontaneous (Ghidoni and O'Neal, 1967) and experimental atherosclerosis (Parker and Odland, 1966a, b; Cookson, 1971). Such 'activated' smooth muscle cells have been identified migrating from the media into the intima of segments of doubly ligated arteries where they have been termed 'myo-intimal' cells (Buck, 1961) (figure 38). It is interesting to note that a similar change occurs in such cells in the aortic

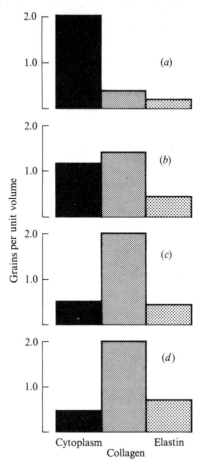

Figure 37. Histograms showing the number of silver grains counted per unit volume of medial component (cell cytoplasm, collagen, elastin) on autoradiographic examination of aortae of one-month-old rats at (a) 15 minutes, (b) 30 minutes, (c) 3 hours and (d) 10 hours after [^{3}H]proline intravenous injection. There is heavy cytoplasmic labelling at 15 minutes which declines steadily over the ensuing period. Label appears later in the collagen and rises as the level of cellular labelling falls. The labelling of elastin is at considerably lower levels but also increases gradually throughout the period of the experiment. Reproduced through the kindness of Dr R. G. Gerrity.

media of aging rats (Cliff, 1970). The interrelationship between the vascular smooth muscle cell and the connective tissue fibroblast has been discussed very fully by Poole and others (1970) and I would agree with their conclusion that these two cells are modulations of a basic cell type and that hard and fast distinctions are not possible. There are other instances where smooth muscle cells are fibroblastic. These are in the hormonally stimulated myometrium (Mark, 1956; Ross and Klebanoff,

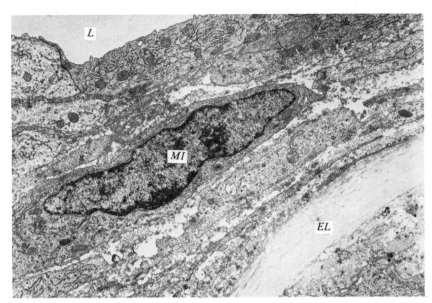

Figure 38. The thickened intima related to the anastomosis of a renal artery in a rabbit. The lumen (L) is lined by plump endothelial cells which are rich in mitochondria and ergastoplasm. The space between the endothelium and the internal elastic lamina (EL) is infiltrated by myo-intimal cells (MI). Magnification × 7250.

1971) and in the distended ureter above an experimental obstruction (Ladányi and Lelkes, 1968).

Apart from the secretory role of these cells in the production of the extracellular components of blood vessel walls there is another specialized secretory adaptation of vascular smooth muscle cells present in the juxtaglomerular apparatus of the kidney (Dunihue, 1967). Here the granular cells in the media of the afferent arteriole and the related 'lacis' cells are derived from vascular smooth muscle cells (Latta, Maunsbach and Cook, 1962; Simpson and Devine, 1966).

No consideration of the functions of vascular smooth muscle cells would be complete without some mention being made of the role of these cells in atherosclerosis and hypertension. The idea has been proposed by a number of investigators that an early essential step in the development of any atherosclerotic plaque is the local weakening and thinning of the tunica media of the vessel consequent on alterations in the smooth muscle cells present therein (Crawford and Levene, 1953; Hass, 1955, 1963; Levene, 1956a, b; Gresham, Howard and King, 1962; Wortman et al., 1966; Zemplényi, 1968). A proliferative intimal fibro-elastic reaction then occurs in relation to the medial defect, which may be looked upon as an adaptive change to re-establish the correct internal diameter of the vessel with respect to its pressure-flow characteristics (Moritz, 1941; Hass, 1963;

La Taillade, Gutstein and Lazzarini-Robertson, 1964; Berry, 1969) or else simply as a reparative or healing response which is confined to the intima (Aschoff, 1933; Baldwin, Taylor and Hass, 1950; Taylor, Baldwin and Hass, 1950; Kelly, Taylor and Hass, 1952; Wortman et al., 1966; Bondjers and Björnheden, 1970). This results in the formation of fibro-elastic intimal plaques that act as the seed-beds for the development of the lipid-filled true atherosclerotic plaques (Vastesaeger and Delcourt, 1962; Daoud et al., 1964). The major cell present within the early intimal proliferative lesions is a smooth muscle cell (Buck, 1962; French et al., 1963; Daoud et al., 1964; Lee et al., 1966; Parker and Odland, 1966a, b; Ghidoni and O'Neal, 1967; Knieriem et al., 1967, 1968; Scott et al., 1967; Still and Dennison, 1967; Becker and Murphy, 1968; Zemplényi, 1968; Lee et al., 1970). With the developing pathology of the atherosclerotic plaque the smooth muscle cells develop lipid inclusions. Evidence both of altered cellular metabolism resulting in lipid synthesis *in situ* (Geer, McGill and Strong, 1961; Whereat, 1967; Cookson, 1971) and of uptake of lipid entering the vessel wall via the endothelium by these cells (Zugibe, 1963; Wissler, 1967) has been found. There is evidence that at least some of the foam cells of the fully developed lesions are derived from the smooth muscle cells (Geer et al., 1961; Parker and Odland, 1966b; Ghidoni and O'Neal, 1967; Scott et al., 1967; Zemplényi, 1968). However, many authorities consider that the foam cells in these lesions arise largely from the monocyte–macrophage series of cells derived from the blood stream (see p. 96).

The basic fault in hypertension is one of imbalance between the propulsive work done by the heart and the resistance supplied by the arterial vessels. The changes that occur in the media of arteries associated with vascular hypertension are illustrated by the work of Aikawa and Koletsky (1970) in the rat. They divided the lesions into four main classes, hypertrophic, hyperplastic, degenerative–fibrotic, and fibrinoid necrotic. Since Crane and Dutta (1963) could identify cellular hyperplasia in the smooth muscle of arteries and arterioles but not of the aorta the increase in extractable muscle protein found in hypertensive rat aortae by Wolinsky (1971) must be attributed to hypertrophic changes. The increase in collagen and elastin found in such vessels (Wolinsky, 1970, 1971) can be associated with increased intracellular organelles within the medial muscle cells (Salgado, 1970) of known fibroblastic capability (p. 86). The adaptive nature of this smooth muscle hypertrophy in the aorta is shown by its complete reversion following the re-establishment of normal blood pressure (Wolinsky, 1971). Increased thickness of medial smooth muscle, whether hypertrophic or hyperplastic in nature, may be of aetiological significance in spontaneous hypertension. Folkow et al. (1970) concluded from their perfusion studies of the hind-limb vasculature of spontaneously hypertensive and control rats that the hypertensive changes were entirely explicable by a 30% increase in thickness of the media in the resistance vessels. This conclusion was supported by further work done on isolated strips of aorta and portal vein from similar animals (Hallbäck, Lundgren and Weiss, 1971). Analysis of wall-to-lumen ratios in cases of human pulmonary hypertension revealed medial muscular hypertrophy in the vessels and also a greatly increased number of small (< 100 μm) arteries in the pulmonary tissues (Wagenvoort, 1960). Samuelson, Becker and Wagenvoort (1970) found that the medial coats of pulmonary veins were very sensitive to changes in pressure, showing hypertrophy and arterialization in response to elevation of pulmonary venous pressure.

Necrosis of smooth muscle cells occurs in the media of the aorta (Salgado, 1970) and of arteries and arterioles (Campbell and Santos-Buch, 1959) of hypertensive animals. Such cellular necrosis is not directly responsible for the accumulation of 'fibrinoid' material within the walls of hypertensive arteries (Moore, Schoenberg and Koletsky, 1963). Fibrinoid material has been identified as a crystalline substance with a 30 nm periodicity in two planes at an angle of 60° (Aikawa and Koletsky, 1970) and as such can be distinguished electron microscopically from fibrin in the walls of arteries (Haust, Wyllie and More, 1965).

STRIATED MUSCLE

To conclude this discussion of muscular elements within the walls of blood vessels brief mention must be made of striated muscle. Cardiac-type muscle may be found within the large veins related to the heart, which apparently possess an intrinsic rhythmicity (Franklin, 1928). Karrer (1960a) identified striated muscle in the walls of veins within the thorax and lungs of the mouse and showed that it had the ultrastructural characteristic of cardiac muscle. In several species of seal voluntary muscle which is innervated by a branch of the right phrenic nerve forms a sphincter controlling the vena cava just anterior to the diaphragm (Walker and Attwood, 1960; Elsner *et al.*, 1971).

THE FIBROBLAST

True connective tissue fibroblasts (Ross and Benditt, 1961; Movat and Fernando, 1962; Cliff, 1963) are present in the walls of blood vessels. They are confined to the tunica adventitia in mammalian vessels (Rhodin, 1962a, 1968; French, 1970) but in birds there is some evidence to suggest that fibroblasts and cells with intermediary forms between fibroblasts and smooth muscle cells are present in the aortic media (Karrer, 1960b; Moss and Benditt, 1970). However, Simpson and Harms (1964) found a pure population of smooth muscle in chick aortic media (see above also in section on smooth muscle). Fibroblasts are more slender than smooth

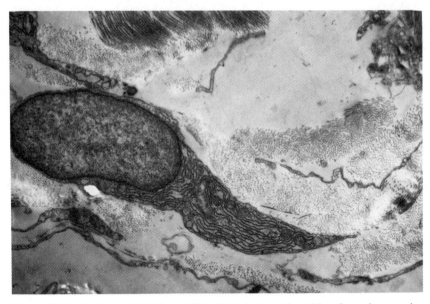

Figure 39. A slender spindle-shaped fibroblast in the adventitia of a rat's aorta has a cytoplasm filled with ergastoplasm and scattered mitochondrial profiles. Slender processes of neighbouring fibroblasts extend between bundles of collagen fibrils related to the cell. Magnification × 7000.

muscle cells, and generally have extremely elongated branching processes which extend between the bundles of extracellular fibrils to come into contact with similar processes from neighbouring cells (figure 39). Regions of membrane nexus have been detected in such regions only in tissue culture (Devis and James, 1962) and in repair tissue (Ryan *et al.*, 1973) and, by contrast with smooth muscle cells, never in normal tissues. The presence of long tenuous cytoplasmic processes tending to enwrap the vessel wall (figure 29) have led to these cells being named 'veil' cells by Rhodin (1968). In the walls of venules fibroblasts can form a continuous coat 0.2 μm thick with individual cells up to 90 μm in length. The cells have an elongated nucleus with generally two prominent nucleoli, large cisternae of rough endoplasmic reticulum (Palade and Porter, 1954) (figure 39), and a well-developed Golgi apparatus. Bundles of intracytoplasmic filaments are sparse in these cells in blood vessel walls. Caveolae are not numerous on the plasma membranes and basement membranes are absent (Rhodin, 1968).

The function of fibroblasts in the adventitia of blood vessels is first and foremost to synthesize and lay down collagen fibrils and ground substances (Jacobson, 1953; Jackson, 1956; Ross and Benditt, 1961, 1962), but presumably they may also be involved in elastic fibre production in sites where they are present within the fibro-elastic laminae (Fahrenbach,

Sandberg and Cleary, 1966). These cells have also been demonstrated to have phagocytic activity within the adventitia of vessels (Bruns and Palade, 1968*b*; Gerrity, 1972) which is not, however, as well developed as that of the macrophage series of cells (below).

Generally speaking, repair processes occurring within the intima and media of blood vessels are performed by smooth muscle cells (p. 87). Repair in the adventitial coat is basically a fibroblastic response. Provided they remain intact, the elastic laminae act as barriers to the ingrowth of fibroblasts and capillaries from the adventitia into the media (Taylor *et al.*, 1950). Hassler (1970) has pointed out that it is only when very strong repair processes occur in the media that some fibroblasts and macrophages may be present. I would presume that defects have been produced by the injurious stimuli in the barrier normally formed by the elastic laminae. Ingrowth of fibroblasts and capillaries from the adventitia, penetrating the media and entering the vascular lumen, occurs in doubly-ligated vascular segments (Schmidt-Diedrichs and Courtice, 1963; Friedman, Byers and St George, 1966). From my own observations on such ligated vascular segments I would deduce that the process is analogous to any other form of granulation tissue ingrowth into a necrosed region. The healing-in of vascular transplants into arteries is accompanied by a fibroblastic and smooth muscle thickening of the intima and a connective tissue proliferative reaction in the adventitia which has a 'splinting' effect on the grafted vessel wall (Watts, 1907; Schloss and Schumacker, 1950).

Whilst certain workers ascribe considerable importance to a population of sub-endothelial fibroblasts in proliferative fibrotic responses of the intima (e.g. Moritz, 1967) the great consensus of opinion is that the intimal smooth muscle cell is by far the most significant cell involved in such lesions (see previous section on smooth muscle). Hence Daoud *et al.* (1964) considered that the 'pre-atheromatous' lesion in the intima involved cellular proliferation of smooth muscle cells only. There appears to be, however, in later stages than this, some admixture of fibroblasts or 'fibroblast-like' cells in the intimal lesions of atherosclerosis (Hass, 1955; Ghidoni and O'Neal, 1967; Scott *et al.*, 1967). It would be a fair assessment, in view of their small numbers and the known fibroblastic capability of the vascular smooth muscle cells present, to state that the fibroblasts are of no great importance in either the initial development or the subsequent progression in pathology of the intimal lesions of atherosclerosis.

PHAGOCYTIC CELLS

In addition to the phagocytic pericytes and the slightly phagocytic fibroblasts there are found in blood vessel walls smaller numbers of highly phagocytic cells of the macrophage–tissue histiocyte-type (figure 40).

The macrophage type of cell (Palade, 1955; Felix and Dalton, 1956;

94 The extra-endothelial cells of blood vessel walls

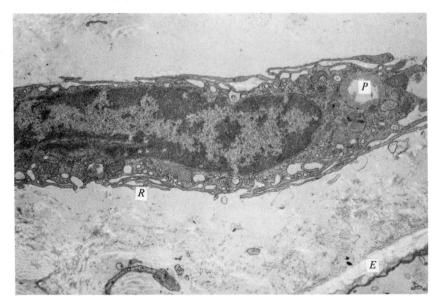

Figure 40. An elongated macrophage in the inner part of the adventitia of the rat's aorta. The margin of the cell is covered with numerous slender veil-like projections or ruffles (R) that are characteristic of this cell. The cytoplasm contains phagosomes (P), numerous vesicles and mitochondria. E, elastic lamina near adventitial–medial border. Magnification × 8000.

Karrer, 1960c; Cliff, 1963) is generally quite large, being 30 μm in diameter, and has a more rounded shape than the other fixed cells of the vessel wall. The cells contain a single large rounded nucleus with a prominent nucleolus. Both rough and smooth elements of endoplasmic reticulum are present in moderate amounts in the cytoplasm, together with free ribosomes and a well-developed Golgi region. There are plentiful rounded and oval mitochondrial profiles scattered throughout the cytoplasm and conspicuous tracts of 10 nm fibrils are frequently present, particularly in relation to the nucleus (De Petris, Karlsbad and Pernis, 1962; Cliff, 1963; Allison, Davies and De Petris, 1971; Uehara et al., 1971) (figure 41). Phagosomes containing ingested material and lysosomes rich in acid hydrolases (Duve, 1959) can be identified within these cells (Cohn and Wiener, 1963; North and Mackaness, 1963; North, 1966b). The plasma membrane has a system of highly developed folds or 'ruffles' extending, sometimes in an extremely complex pattern, out from the cell surface (Palade, 1955; Felix and Dalton, 1956; Karrer, 1960c). It is these structures, as identified in the electron microscope, that correspond to the cytoplasmic 'veils' of macrophages seen *in vivo* and *in vitro* (Lewis, 1925–6), which are responsible for the trapping and engulfment of material in their locality (figure 42). The presence of these structures,

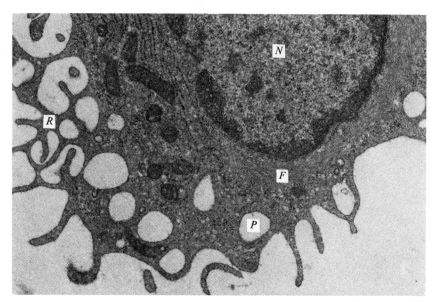

Figure 41. A macrophage in an exudate in a rabbit's pericardium. The peripheral cytoplasm is thrown into a complex series of branching and fusing ruffles (R) whose activity gives rise to numerous phagocytic vesicles (P). A well-developed tract of fine intracytoplasmic fibrils (F) partially surrounds the nucleus (N). Elsewhere the cytoplasm contains profiles of ergastoplasm and mitochondria. Magnification × 22 500.

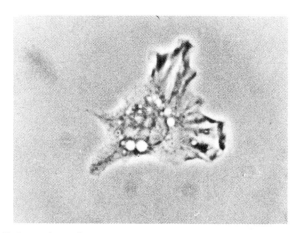

Figure 42. Guinea pig peritoneal macrophage *in vitro*. Note the ruffled peripheral cytoplasm and the numerous bright cytoplasmic vacuoles associated with the phagocytic activity of these cells. Phase optics, magnification × 1100.

96 The extra-endothelial cells of blood vessel walls

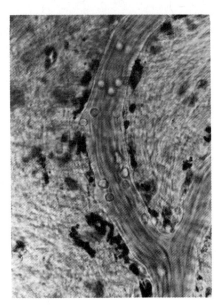

Figure 43. Numerous black carbon-containing macrophages situated along the walls of a branching venule within a rabbit ear chamber. In-vivo photomicrograph, magnification × 320.

together with the demonstration of lysosomal enzymes (usually acid phosphatase) and phagosomes, are the most reliable criteria for the identification of macrophages.

Macrophages are normally present in small numbers within the adventitia of the larger vessels, generally related to the vasa vasorum. In the vessels of the microcirculation macrophages occur very constantly in relation to the vessel walls (Ebert and Florey, 1939; Cotran et al., 1965), most prominently those of venules (figures 5, 43). The important association of blood vessels and specialized phagocytic cells within the reticulo–endothelial system has been described above.

The 'foam cells' of atheromatous deposits are derived partially from smooth muscle cells (p. 90) and also from macrophage-type cells or 'lipophages' (Cookson, 1971). These cells are derived from circulating blood monocytes (Still, 1964; Marshall and O'Neal, 1966) which may contain lipid inclusions and be identified as macrophages (Hass, 1955; Poole and Florey, 1958; Still and O'Neal, 1962; Hüttner, More and Rona, 1970). They enter the intima apparently by way of the overlying endothelial membrane. Foam cells within the intima of arteries are capable of proliferation as shown by [^{3}H]thymidine incorporation studies (Spraragen, Bond and Dahl, 1962). Macrophages do not normally occur within the media of vessels but may be found there in very strong repair processes (Hassler, 1970).

5
THE EXTRACELLULAR COMPONENTS OF BLOOD VESSEL WALLS

THE BASEMENT MEMBRANE SYSTEMS

The general distribution of these structures in relation to pericytes and smooth muscle cells has been dealt with in the sections devoted to each of these cell types. Practically all endothelial cells of blood vessels possess basement membranes related to their abluminal surfaces (Rhodin, 1962a). However, there is considerable variation in the completeness of the basement membrane from one type of vessel to another (Bennett et al., 1959) and also in the difficulty of obtaining adequate fixation of this structure for electron microscopy. The vessels of the earthworm are reported as having their basement membranes related to the luminal rather than the abluminal endothelial surfaces (Hama, 1960).

Histologically, the basement membrane region corresponds to a layer of glycoprotein demonstrable by the periodic acid–Schiff (PAS) (Pearse, 1960) and Hotchkiss techniques (Gersh and Catchpole, 1949) and appears to be a condensation of the ground substance of the surrounding tissues. In general, basement membranes exist wherever connective tissue forms a boundary (Jacobson, 1953). Immunohistological techniques have shown that they contain collagen (Rothbard and Watson, 1961). Majno (1965) considered that these structures are largely composed of non-banded collagen, in either a fibrillar or amorphous form, which is susceptible to treatment with collagenase.

Electron microscopically the endothelial basement membrane consists of a layer of fibrillar material (10 nm diameter) condensed into a feltwork (Palade, 1953a) within a faintly granular or amorphous continuum (Pease and Molinari, 1960; Latta, 1970; Leak, 1971). The fibrillar material is very similar to the fine fibrils scattered throughout the connective tissue ground substance (Majno, 1970). The endothelial basement membrane is always separated from the related plasma membrane of the cell by a clear zone about 25 nm wide (Policard and Collet, 1958; Pappas and Tennyson, 1962; Leak, 1971) (figures 14, 16). The condensations are of variable thickness, having been measured as 50–60 nm (Fawcett, 1959), 100 nm (Rhodin, 1962a) and 150 nm (Latta, 1970). Careful studies of these structures by Latta (1970) and Policard et al. (1957) reveal the presence of three layers within them but these two reports differ as to the relative densities of the layers. Latta found the central layer to be the denser whilst Policard and others found this layer to be lighter than the outer layers. Usually the basement membrane is present as a single condensation, but at times it may be reduplicated or even laminated in appearance (Hogan and Feeney, 1963; Wolff, 1963; Ludatscher and Stehbens, 1968). As described earlier, the endothelial basement membrane can merge with the basement membranes of related pericytes and smooth muscle cells or be absent in regions where endothelial projections form cell-to-cell contacts with such underlying cells. In certain vessels, such as the aorta (Pease and Paule, 1960; Gerrity and Cliff, 1972) and the coronary arteries (Parker, 1958), endothelial basement membranes are difficult to demonstrate, but as shown by the work of Ts'ao and Glagov (1970a, b) they can in fact be demonstrated in the aorta if suitable techniques are employed.

Fawcett (1959) considered that the vascular basement membrane was a product of the endothelium alone. Hadfield (1951), in studying new vessel formation in granulation tissue, considered that the basement membrane was formed first as a saccular outgrowth into which a lining of endothelial cells subsequently migrated. From my own observations on new vessel formation during wound healing (Cliff, 1963) it would appear that new basement membranes are formed in relation to the invading endothelial 'sprouts', very near to, but not at, their tips. It is my belief that these condensations are an expression of some interaction between the endothelium and the connective tissue and not solely a product of the endothelial cells. The absence of basement membrane at the invading tips of the sprouts is in agreement with Schoefl (1963) who found that endothelial invasion occurred in relation to defects in the basement membranes. The findings of Taylor, Shepherd and Robertson (1961), which showed the binding of fluorescein conjugated anti-reticulin antibodies to both the capillary basement membrane and the capillary endothelial cytoplasm of newly formed vessels, support the concept that the endothelium contributes some component to the basement membrane.

Endothelial basement membranes perform several functions within vessel walls. They provide mechanical support to the vascular endothelium (Policard and Collet, 1958) and serve to anchor securely the endothelium at various points where tracts of fine fibrils extend from the cells to merge with the basement membrane (Policard et al., 1957; Ts'ao and Glagov, 1970a). The mechanical resistance supplied by this layer can be appreciated by realizing that it retains leukocytes and particles that have traversed the endothelial membrane. This occurs with granulocytes (Marchesi and Florey, 1960), chylomicra (Majno and Palade, 1961; Schoefl and French, 1968) and colloidal tracer particles (Majno and Palade, 1961; Marchesi, 1962; Hurley, 1964; Cotran et al., 1965) in the inflammatory reaction (figures 27, 44). The escape of such accumulated elements apparently depends upon the ability of the emigrated leukocytes to

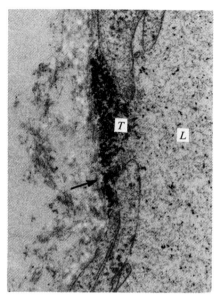

Figure 44. An endothelial gap within a highly permeable vessel of rabbit granulation tissue containing numerous particles of intravenously injected ThO_2 label (T). These particles are being retained by the endothelial basement membrane (arrow). L, lumen. Magnification ×21 000.

produce local defects within the restraining basement membrane. Only slight hold-back of lymphocytes occurs between the endothelium and the basement membrane of the specialized 'adenoidal' vessels of lymphoid tissues (Claesson, Jørgensen and Røpke, 1971; Schoefl, personal communication) (figure 45). Basement membranes also act to hold up, to some extent, the physiological diffusion of materials leaving the blood vessels. Hence Jennings and Florey (1967), using a perfused heart preparation, found that particles of saccharated iron oxide were retained by capillary basement membranes. However, in general, it would appear that it is only with particles of 10 nm diameter, or larger, that a true diffusion barrier can be identified (Farquhar, 1960; Latta, 1970). These structures can therefore play only a minor role in the control of permeability, and Cotran and others (1965) and Karnovsky (1967) found that particles smaller than colloidal carbon passed through them easily while carbon was only temporarily held back.

The ground substance as a whole, with its content of charged groups bound to polysaccharide molecules, seems likely to be the stationary phase of an ion-exchange system (Bennett, 1963). Gersh and Catchpole (1949) considered capillary permeability to be linked inseparably to the physical state of the ground substance surrounding the blood vessels, and recently Stearner and Sanderson (1971) have found evidence of structural

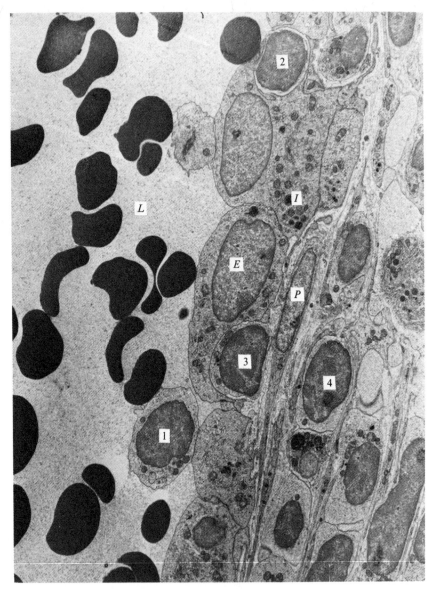

Figure 45. The wall of a lymphocyte-secreting ('adenoidal') venule from a Peyer's patch of a rat. Lymphocyte 1 is still within the lumen (L) of the venule and attached to the high endothelium (E) which contains pleomorphic dense inclusions (I). Lymphocyte 2 is passing down an endothelial junction. Lymphocyte 3 is trapped between the endothelium and its basement membrane. A large number of lymphocytes are lined up beside the vessel wall (4). P, pericyte. Magnification × 5000. Electron micrograph supplied through the kindness of Dr G. I. Schoefl.

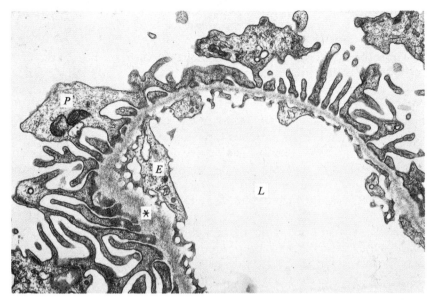

Figure 46. Part of the glomerular tuft from a rabbit kidney fixed by intra-arterial infusion. A fenestrated endothelial sheet (E) lines the lumen (L). The podocytes (P) are separated by a dense basement membrane which in part of the figure is obliquely sectioned (∗). Magnification ×16000.

organization within the ground substance that may be of importance in controlling blood–tissue exchanges. Highly permeable vessels, such as occur in frogs (Stehbens, 1965), the ciliary body of the eye and choroid plexus (Pappas and Tennyson, 1962) and, it now appears, in the hepatic sinusoids (Grubb and Jones, 1971), all have well-developed basement membrane systems. In the renal glomerulus true filtration occurs at the podocyte slits (Graham and Karnovsky, 1966; Latta, 1970) and the basement membrane system plays no significant role in this regard. Relatively impermeable vessels rely upon the endothelial barrier for control of permeability (Bodenheimer and Brightman, 1968; Giacomelli et al., 1970).

The renal glomerular basement membrane has been studied more intensively than any other in the body. It is a very thick (∼150 nm), three-layered structure (Latta, 1970) that separates the endothelium of the vessels from the epithelial cells, or podocytes, of the glomerulus (figure 46). Kurtz and Feldman (1962) performed experiments with argyria in rats and considered that the development of a subepithelial zone of basement membrane 'unstained' by silver deposition after stopping the administration of silver nitrate in the drinking water was proof that the podocytes were responsible for the production of the basement membrane. Recently Striker and Smuckler (1970) repeated these experiments and questioned whether the clear subepithelial zone that appears on ceasing silver nitrate treatment in fact represented synthesis of normal glomerular basement membrane.

Whilst the actual source of this abundant basement membrane material is, therefore, still not definitely established, Farquhar and Palade (1962) found evidence of a continuous uptake of this material by the related podocytes. They considered this to be part of a renewal system of the basement membrane. Further evidence for the turnover of the materials forming the glomerular basement membrane has been provided by McPhaul and Dixon (1969). They detected two cross-reacting carbohydrate-containing antigens of the glomerular basement membrane within both the blood and the urine of normal people.

Vascular basement membranes show both a generalized diffuse and a segmental form of thickening with aging in the vessels below the level of the heart (Williamson, Vogler and Kilo, 1971). The severity of the changes they observed were proportional to the distance below the heart in both man and in the legs of the giraffe. However, a purely gravitational cause is difficult to prove in light of the failure by these investigators to demonstrate similar changes in the vessels from different levels in the giraffe's neck between which there existed up to 8–10 feet difference in height.

COLLAGEN

This is one of the most ubiquitous proteins of the animal kingdom (Randall et al., 1952). In vertebrates one third of the total mass of body protein is comprised of collagen (Marbach, 1957; Udenfriend, 1966). Its function in tissues is to provide mechanical strength by virtue of its fibrillar nature and high tensile strength (Glagov and Wolinsky, 1963; Wiederhielm, 1965; Dunihue, 1967). It may be readily recognized histologically by its characteristic staining properties, particularly in the various trichrome preparations (Pearse, 1960). Collagen fibrils are anisotropic when examined by polarized light and this feature is useful for their identification both in sectioned material and in living tissues. They do not show autofluorescence when examined by ultraviolet light. Certain collagen fibres may be associated with increased amounts of polysaccharide material which causes them to be stained by silver techniques and the PAS reaction. Such argyrophil fibres, also known as reticulin fibres, merge with normal collagen fibres in tissues (Jacobson, 1953). It is considered that such fibres represent a particular stage in collagen deposition (Maximow, 1928; Jackson, 1957; Williams, 1957). Electron microscopic examination of argyrophil fibres has shown individual collagen fibrils (Kramer and Little, 1953) embedded within coats of argyrophilic material (Wasserman and Kubota, 1956). Reticulin and basement membranes are antigenically very similar (Cruikshank and Hill, 1953; Taylor, Shepherd and Robertson, 1961). Anti-collagen antibodies, however, are unable to combine with the antigenic groups present within reticular fibres (Mancini et al., 1965). Reticulins and basement membranes share common antigens that recently formed argyrophilic fibres do not possess. Hence, it is possible to separate argyrophil fibres into two categories. The first represent a stage in the deposition and maturation of normal collagen fibres whilst the second represent the true reticulin fibres of tissues. The latter fibres are permanent features, representing definite structural entities, and they do not mature to take on the tinctorial properties of collagen. Basement membranes, whilst sharing tinctorial and immunologic features with reticulins, can be clearly distinguished by their ultrastructure (see p. 97).

An important association between reticular fibres and endothelial cells is found in the reticulo-endothelial system. The intimate associations that are found between endothelium and reticular fibres, as in this and other instances, has led to claims that endothelial cells secrete these fibres (Corner, 1920).

Histologically, collagen appears to be composed of fibres with diameters of 0.3–0.5 μm (Jacobson, 1953). Using the electron microscope these fibres are found in fact to be bundles of very much finer fibrils. According to Jackson (1956) the first recognizable collagen fibrils have diameters of 8.0 nm which increase with maturity to an average diameter of 40.0 nm. Individual collagen fibrils have a highly characteristic banding along their lengths with a major periodicity of 64.0 nm (Schmitt and Gross, 1948; Grant, Horne and Cox, 1965) (figure 47). The regular pattern of sub-banding within the fibrils can be explained in terms of various stacking arrangements of the individual 280 nm tropocollagen molecules composing the fibrils (Hall and Doty, 1958; Olsen, 1963; Grant et al., 1965). The anisotropy of collagen is due to the high degree of ordering of this basic protein structure (Picken, 1940). Collagen has a highly characteristic amino acid composition (Piez, Weiss and Lewis, 1960) with high contents of non-polar amino acids such as glycine, proline and hydroxyproline. The role of fibroblasts and smooth muscle cells in its elaboration has been discussed in chapter 4.

The major site of deposition of collagen in the walls of blood vessels is the tunica adventitia of arteries and veins where it is present with a variable admixture of elastic fibres (French, 1970). It has been proposed that the adventitia acts as the main mechanical constraint to stresses tending to distend the vessel wall (Rhodin, 1962a; Dunihue, 1967). Since collagen fibrils within the adventitia are aligned mainly in the long axis of the vessel, with probably a coarse helical arrangement of individual bundles (Wiederhielm, 1965; Gerrity, 1972), they would obviously act to resist longitudinal stresses considerably better than radial or tangential stresses. This is supported by measurements made of the relative distensibility of

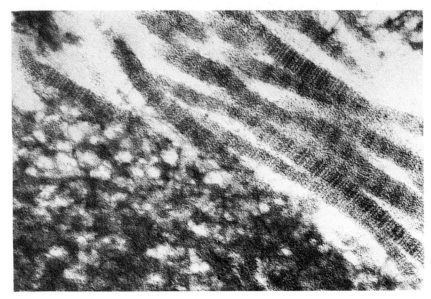

Figure 47. Collagen fibres (upper right) and elastica (lower left) run diagonally across the figure. The collagen fibres show characteristic period and sub-banding and the elastica appears as a meshwork of dense fibrillar material. Rat aortic media. Magnification ×120000.

arterial walls which indicate that they are less distensible in the longitudinal than the radial direction (Burton, 1967). Peterson (1962) estimated the length changes in large arteries during the cardiac cycle to be about 1%, with concomitant changes occurring in their diameters of up to 4%. Gerrity (1972), using in-situ perfusion fixation at physiological pressure, found a high degree of order and cross-linking of adventitial collagen fibrils, analogous to the organization present within tendons.

Collagen is morphologically plentiful within the large elastic arteries (Karrer, 1961; Glagov and Wolinsky, 1963; Paule, 1963; Wolinsky and Glagov, 1964, 1967a; Städeli, 1966; Cliff, 1970; Moss and Benditt, 1970). By chemical analysis, it makes up from 16% (pig) to 25.6% (rat) of the dry fat-free weight of the aortic arch (Neuman and Logan, 1950). Gerrity (1972) investigated the collagen content of the rat's thoracic aortic media by morphometry, to give a percentage by volume, and chemically, to give a percentage dry weight, and found that in the newborn animal the collagen content was 2.7% by volume and 8% by weight, rising to 19% by volume and 23% by weight at two months old. At six months of age the collagen content had increased still further to about 30% (by weight). The amount of collagen present within the aortic media varies from species to species and rises progressively from arch to iliac bifurcation within each animal (Grant, 1967). As well as showing a longitudinal gradient,

Feldman and Glagov (1971) were able to demonstrate that the aortic media also shows a radial gradient in collagen distribution. These authors found in human autopsy material that in children the aortic medial collagen content increased progressively in amount as the wall was traversed from the intimal to the adventitial surface. They found that this pattern was reversed in old age with a fall in collagen content from intimal to adventitial surface of the media, whilst in young adults they could detect no gradient across the media.

Collagen is present within the media of muscular arteries and of arterioles (Rhodin, 1962a) but is only present in significant quantities, morphologically, in the large muscular arteries (Pease and Molinari, 1960). Veins contain considerably more connective tissue in their walls than arteries of equivalent size, which would account for their greater internal bursting pressure as compared to the arteries (Franklin, 1928).

The collagen fibres of the media of blood vessels are extremely important in determining the physical characteristics of the vessel as a whole. These fibres bear the major part of the stressing forces operating on the walls of arteries (Glagov and Wolinsky, 1963; Wolinsky and Glagov, 1964). Burton (1967) proposed that the feature exhibited by arteries, whereby the more they are stretched then the greater they resist further stretching (Bergel, 1961), was in large measure determined by the 'slack' being taken up in progressively larger numbers of collagen fibres, with the final shape of the tension–length curve being determined by the number and the strength of the collagen fibrils present. Levene (1961) found a good correlation between increasing collagen content and increasing breaking strength of the aorta of the developing chick. The collagen of the media is disposed mainly circumferentially and is thus aligned to resist tangential stresses acting on the vessel wall.

Wiederhielm (1965), working on small arteries (60–150 μm diameter) of the frog mesentery, arrived at similar conclusions. Using planimetric measurements from electron micrographs to estimate the relative proportions of the components of the vessel intima and media, collagen accounted for $21 \pm 0.8\%$ ($n = 20$) by volume, and by computation it contributed 63×10^6 dynes cm^{-2} versus 0.32×10^6 dynes cm^{-2} for the remaining constituents (endothelium, internal elastic lamina and smooth muscle) to the stiffness of the wall. He found that 'slack' collagen fibres progressively took strain as the vessel wall distended but that even at small degrees of strain ($\sim 5\%$) at least 80% of the wall stress was supported by collagen.

Collagen is normally present within the tunica intima of the larger vessels but shows considerable variation in amount from species to species (French, 1970). The human aorta has a subendothelial fibro-elastic zone that contributes one sixth of the wall thickness. Levene and Poole (1962) found that this tissue was composed of $\sim 20\%$ collagen on a dry weight basis, and since it was mainly laid down in the first 2 decades of life they considered that it definitely had a physiological function in the intima. Intimal collagen fibres can be identified in the aortae of the smaller mammals generally used for laboratory experimentation such as the guinea pig (Gore, Fujinami and Shirahama, 1965), the rabbit (Ts'ao and Glagov, 1970*b*), the rat (Gerrity and Cliff, 1972) and the mouse (Karrer, 1961).

The identification of collagen fibrils in the intima of arterial vessels may be of considerable importance in the reactions of the blood platelets that occur following endothelial denudation. Caen, Legrand and Sultan (1970) considered that collagen–platelet interactions were of great significance in initiating platelet adhesion, aggregation and degranulation. In studies of the fine structure of platelets in contact with intimal tissues denuded of endothelium, platelets have been identified adhering to both collagen fibrils and the fine microfibrils of connective tissues (Ts'ao and Glagov, 1970*b*). Stemerman, Baumgartner and Spaet (1971), on the other hand, found in similar experiments that platelets only adhered to the fine extracellular microfibrils of the intima, which were distinct from collagenous material and were probably related to the microfibrils of elastic tissue. Platelet adhesion to denuded basement membrane surfaces (Warren and De Bono, 1970) could also be explained in terms of the collagenous components of these structures (Caen *et al.*, 1970). Platelet adhesion to the walls of arteries, with subsequent thrombus formation and organization, has been suggested as a source of atheromatous plaques (Crawford and Levene, 1953; Duguid, 1954). The relation between possible endothelial damage and denudation and altered boundary layer flow conditions has been discussed previously (chapter 2).

The amount of collagen present within the walls of arteries shows a general increase with advancing age. This fibrosis, or *sclerosis*, of arterial walls is associated with both dilatation of the lumen and overall lengthening of the vessels (Aschoff, 1933) which leads to tortuosity and, in extreme cases, marked spiralling of arteries (Hassler, 1969) (figure 48). The physical properties of the arterial walls alter with age, showing a decreased volume distensibility and an increased pulse wave velocity (Burton, 1967). Peterson (1962) has drawn attention to the difference between vessel wall distensibility and vessel wall stiffness and has pointed out that as vessels become stiffer with age, their distensibility does not decrease to the same degree. Learoyd and Taylor (1966) have pointed out that when vessel wall thickness and radius are taken into account it becomes apparent that the tissues of the walls at all sites in the vascular tree become weaker, in absolute terms, with advancing age. A good deal of work has been devoted to studying the progressive fibrosis of the aortic wall with aging. Increased collagen is biochemically detectable both in the intima (Levene and Poole, 1962) and the aortic wall in general (Kao and McGavack, 1959; Fontaine *et al.*, 1968). Fontaine and others have established that the

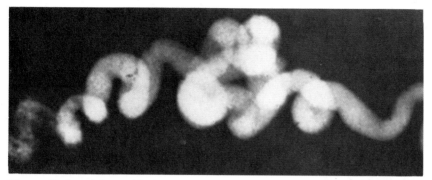

Figure 48. Arterial vessel in human sciatic nerve showing senile alteration to form glomerular loops. Micro-angiogram, magnification ×110. (Hassler, 1969.)

progressive increase in collagen is not due to increased rate of synthesis but due to decreased rate of turnover of this protein, which also shows an altering amino acid composition with age. My own observations on the aortae of aging rats have revealed the presence of extremely wide atypical collagen fibrils which may well be a morphological expression of progressive lateral aggregation of collagen associated with such a depressed turnover (figure 49). Collagen deposition with aging can be detected histologically in the aortic media of human beings (Hass, 1943) where it occurs both diffusely and also as roughly triangular fibrous 'splints' extending from each surface of the elastic laminae. Increased collagen associated with aging has also been detected histologically in various laboratory animals – in the guinea pig (Städeli, 1966), the rat (Cliff, 1970) and the mouse (Smith, Seitner and Wang, 1951; Karrer, 1961). Such aging changes in the human aorta led Urschel *et al.* (1968) to investigate the effects upon the myocardium of routing the cardiac output through a non-distensible stainless steel tube that by-passed almost the entire length of the descending aorta. They found that this resulted in increased impedance to cardiac ejection and an increased tension load on the myocardium However, as pointed out by Hallock and Benson (1937), the aorta of old age functions as a capacity chamber, or reservoir, which is able to accept the cardiac output without placing undue strain upon the heart, due to the increased length and diameter of such sclerosed vessels. Learyod and Taylor (1966) investigated the changes associated with aging in vessels at all sites in the arterial tree and considered that they resulted in overall beneficial effects in maintaining non-uniform impedance characteristics, showing an increase towards the periphery of the tree, with attendant haemodynamic advantages.

In atherosclerotic lesions of blood vessels there is invariably found an association of a basically collagenous fibrotic proliferation with a lipid infiltration (Duguid, 1954; Ghidoni and O'Neal, 1967) that may be both

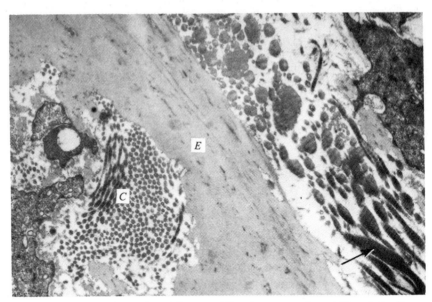

Figure 49. An elastic lamina (*E*) runs obliquely across the middle of the figure. Extremely wide collagen fibres with typical cross-striations (arrow) are present to the right of the lamina. More normal appearing collagen is present to the left of the lamina (*C*). Thoracic aorta of a food-restricted rat. Magnification × 20000. Tissue supplied by Dr A. V. Everitt.

intra- and extracellular (McCombs, Zook and McGandy, 1969). There is considerable debate as to which of these two components is the initiating or prime cause in this most important disease. One school considers that initially lipid accumulates within foam cells in the intima which provoke a fibrotic reaction (Anitschkow, 1967). A fibrotic reaction is also associated with the death of foam cells and the release of their contained lipid into the extracellular space (Geer, McGill and Strong, 1961; Zemplényi, 1968). Spain and Aristizabal (1962) have shown that both cholesterol and cholesterol esters, which form a significant proportion of the lipid component of atheromatous plaques (Adams, Bayliss and Ibrahim, 1963; Hass, 1963; Hollander, 1967; Weller, 1967; Whereat, 1967), were the most potent lipids in evoking inflammatory and fibrotic reactions in connective tissue. Adams *et al.* (1963) also examined the sclerotic reaction to cholesterol and considered that phospholipids had an ameliorating effect in promoting phagocytosis of the lipid with consequent resolution of the lesion with little or no resulting fibrosis.

There is another school that considers that the initial intimal lesion within which lipids subsequently accumulate is fibrotic (Gresham, Howard and King, 1962; Vastesaeger and Delcourt, 1962; Daoud *et al.*, 1964). In addition, some workers consider that the intimal lesions of atherosclerosis develop in areas overlying regions of local degeneration and fibrosis, within the media (Paterson, Slinger and Gartley, 1948; Lansing, 1961). Duguid (1954) maintained that any degenerative process within the arterial wall would, of necessity, lead to dilatation with widening of the lumen and not to the narrowing which is associated with the atherosclerotic process. The observations of Crawford and Levene (1953) have shown that atheromatous plaques overlie regions of medial thinning. They considered that in effect a shallow aneurysm, produced by local weakness of the media, was the initial lesion and that the atheromatous plaque was due to mural thrombotic accretion and organization. I would agree more with the view of Hass (1955, 1963), expressed previously, that the intimal reaction is basically an adaptive and beneficial form of proliferation in response to altered pressure-flow conditions produced by local dilatation. Direct confirmation of the association between dilatation of the aorta with the formation of local intimal plaques has been obtained in rabbits (La Taillade et al., 1964).

Repair processes in the walls of vessels are very similar to repair histogenesis elsewhere in the body (Gillman, Hathorn and Penn, 1957) which result in the formation of scar tissue rich in collagen (Haimovici, Maier and Strauss, 1958; Merkow, Lalich and Angevine, 1961; Murray, Schrodt and Berg, 1966; Gerrity and Cliff, 1972). Björkerud (1969) compared the repair processes occurring in longitudinally and transversely oriented intimal injuries in the aorta and found that longitudinal lesions healed to leave no residual intimal thickening whilst transverse ones did not remodel and left a persistent intimal plaque which could subsequently become infiltrated by lipids.

Increased mural tension, as occurs in hypertension, leads to increased absolute amounts of collagen within the walls of arteries (Wolinsky, 1970, 1971). By way of contrast it is interesting to note that a fall in collagen content occurs within the walls of actual and 'potential' varicose veins (Švejcar et al., 1964).

Accumulation of mucopolysaccharide material has been observed to precede intimal fibrotic lesions in arteries (Gillman et al., 1957; Wexler, 1964; Wexler, Judd and Kittinger, 1964; Berlepsch, 1970). Kaplan and Meyer (1960) studied the mucopolysaccharides of the human aorta and found that the total yield showed no correlation with age but that chondroitin sulphate-B and heparitin sulphate both increased whilst hyaluronate and chondroitin sulphate-C decreased their proportions of the total with age and with increasing severity of atheroma. These changes were considered more complex than those which occur in other aging tissues.

ELASTICA

There is probably more confusion in the minds of histologists, physiologists, biochemists and pathologists concerning the properties of this unique substance than those of any other of the components of blood vessel walls. Part of the problem stems from the use of terminology in which elastic tissue, or *elastica*, is frequently equated with *elastin*. Elastin is the protein constituent of elastic tissue which can be extracted by chemical means from tissues (Partridge and Davis, 1955; Jackson and Cleary, 1967). This involves the removal of cellular materials, ground substance and collagen, generally by treatment of the tissue with techniques involving autoclaving and hot alkali extraction (Gotte and Serafini-Fracassini, 1963; Serafini-Fracassini and Tristram, 1966), but also at times employing acetic acid (Hall, Reed and Tunbridge, 1955), peracetic acid (Cox and Little, 1961) or formic acid (Cox and O'Dell, 1966). Such techniques have been quite justifiably termed harsh by Bentley and Hanson (1969). What remains after such extraction is identified as the elastin of the tissue. Such material in its hydrated state is a sticky, yellow–brown material with properties reminiscent of uncured rubber. The amino acid composition of elastin has been determined (Lansing, 1954, 1959; Partridge and Davis, 1955; Serafini-Fracassini and Tristram, 1966). It contains hydroxyproline, estimated as between 2% (Stetten, 1949) and 2–4% (Bentley and Hanson, 1969) of the total amino acids, which is a feature shared only with collagen. The high proportion of non-polar amino acids in the elastin molecules has led to a globular form of structural unit being proposed (Partridge, 1966). Weis-Fogh and Andersen (1970) on microcalorimetric investigation of stretched hydrated elastin obtained results consistent with a model wherein globular particles were deformed and increased their surface areas. Since there are insufficient hydrophilic (polar) groups to cover the elastin molecules (mol. wt. 67000) additional hydrophobic groups would be exposed at the surface, causing a reduction in entropy with consequent heat liberation. They proposed that such a model be termed a liquid drop elastomer.

The most specific chemical feature of elastin is the presence of desmosine and iso-desmosine, which are multiple α-amino acids produced during elastogenesis by the condensation of four lysine molecules into a ring structure (Partridge, Elsden and Thomas, and Dorfman, Telser and Ho, 1964). The fact that desmosines exhibit blue-white autofluorescence may well account for the autofluorescence exhibited by elastic fibres (Partridge, 1962) and purified elastin (figure 50). These condensations are thought to be sites of cross-linkage between the macromolecules of elastin. Serafini-Fracassini and Tristram (1966) compared the calculated distance separating desmosine linkages, with a frequency of three residues per 10^5 g of elastin, with the distance separating 'Y' junctions which were observed

Elastica 111

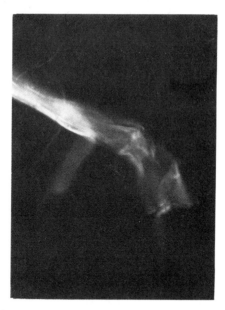

Figure 50. A fibril of reconstituted elastin prepared chemically from calf ligamentum nuchae showing autofluorescence when being examined with ultraviolet light. Magnification × 300.

electron microscopically in the fibrils present within ultrasonically disrupted aortic elastin. The theoretical distance was 110 nm and the observed distance was 130 nm.

These results, indicating that the protein component of elastic tissue may be present in a fibrillar form, find a large measure of agreement in the literature, although considerable variation exists in the actual diameters of the basic filaments observed by various workers. Hence fibril diameters of 35 nm (Schwarz and Dettmer, 1953), 25 nm (Lansing *et al.*, 1952), 20 nm (Dempsey and Lansing, 1954; Hall *et al.*, 1955), 7.0 nm (Gross, 1949; Rhodin and Dalhamn, 1955), 4.4 nm (Cliff, 1971), 3.0 nm (Gotte, Meneghelli and Castellani, 1965; Fahrenbach *et al.*, 1966) 2.0 nm (Cliff, 1971) and finally 1.5 nm (Gotte and Serafini-Fracassini, 1963; Simpson and Harms, 1964) have been reported. The variation reflects, to a large measure, differences in techniques. The lower values obtained accord well with the dimensions of other fibrillar proteins such as α-keratin (2.0 nm) (Dobb, 1964, 1966), silk fibroin (~ 2.4 nm) (Dobb, Fraser and Macrae, 1967) and collagen (~ 1.5 nm) (Hall and Doty, 1958; Tromans *et al.*, 1963). The disposition of the fibrils within both elastic tissue and elastin in a loose network of randomly oriented and tangled filaments (Cox and Little, 1961; Partridge, 1962; Cliff, 1971) (figures 33, 47) agrees with the theoretical structure of elastomers

(King and Lawton, 1950; Partridge, 1962). It should be remembered that elastin requires hydration, which acts to provide a lubricant, to show elasticity (Fahrenbach et al., 1966).

The protein fibrils of elastin are embedded within an amorphous matrix material (Gross, 1949; Lansing et al., 1952; Schwarz and Dettmer, 1953; Hall et al., 1955; Partridge, 1962; Gotte and Serafini-Fracassini, 1963; Simpson and Harms, 1964; Gotte et al., 1965; Fahrenbach et al., 1966; Cliff, 1971) which has the important functions of providing bulk to the elastic fibre and of absorbing thermal energy liberated during stretching (King and Lawton, 1950), and also of acting as a lubricant allowing free movement of the fibrils embedded within it during lengthening and shortening. The amorphous matrix is generally considered to have the same chemical composition as the fibrillar component (Lansing, 1959). Gotte and others (1965) considered, however, that human aortic elastin consisted of fine fibrils embedded within a second protein that was selectively removed by treatment with hot alkali and which, in contrast to elastin as a whole, had large numbers of polar side chains. Gross (1949) observed that trypsin attacked the amorphous matrix material to reveal filamentous structures, but subsequently Franchi and De Robertis (1951) showed that no filaments were produced by this procedure if adequate sterility was maintained, indicating a probable bacterial origin for the filamentous structures. Hall and others (1955) found the matrix material to be unaffected by trypsin but to be removed through the action of pancreatic elastase (Baló and Banga, 1950) to reveal filaments which were slowly broken down by the continuing action of the elastase. It appears that elastase has a non-specific proteolytic action (Partridge and Davis, 1955; Partridge, 1962). I have been able to demonstrate a biphasic reaction of the amorphous matrix material on treatment with osmium tetroxide (OsO_4) (Cliff, 1971). This component is fixed histologically within aortic tissue and purified elastica by treatment with OsO_4 for periods up to twenty-four hours. Prolonging this treatment to forty-eight hours and longer results in selective removal of the matrix material to reveal the underlying fibrillar sponge-like structure of the elastic tissue (figures 33, 47).

The combination of randomly oriented fibrils and amorphous matrix material identified electron microscopically is supported by the results of X-ray diffraction studies when elastin gives the pattern expected of a non-crystalline polymer (Lansing, 1959; Cox and Little, 1961). This lack of internal ordering is also confirmed by examination of elastic fibres with polarized light. Such fibres are normally isotropic but on stretching can become anisotropic due to alignment of the filaments (Dempsey and Lansing, 1954).

Elastic fibres (or elastica) can be readily recognized as yellow, wavy, branched refractile fibres, which in blood vessel walls are commonly

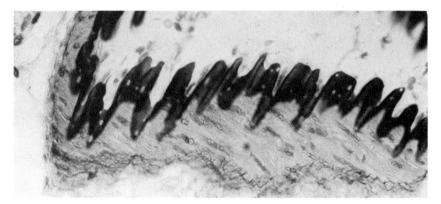

Figure 51. The darkly-stained internal elastic lamina of a rat lumbar artery thrown into numerous folds and ridges. Clear holes, or fenestrae, can be seen scattered randomly in the membrane. Verhoeff–Van Gieson stain. Magnification × 480.

condensed into fenestrated lamellae (Dempsey, 1952; Parker, 1958; Partridge, 1962; Cliff, 1970) (figure 51). Elastica stains with eosin and gives a faint positive reaction to the PAS technique (Dempsey and Lansing, 1954), not necessarily denoting the presence of polysaccharide material (Pearse, 1960). There are a number of relatively specific staining reactions that are extremely valuable for demonstrating elastica (see Pearse, 1960). I have found the Verhoeff stain used in combination with Van Gieson's stain to be particularly useful in examining blood vessel walls histologically. The fact that elastic fibres can be demonstrated histochemically to contain lipid material (Dempsey and Lansing, 1954) finds confirmation in the observation of the release of oily droplets from elastica treated with elastase (Lansing et al., 1952). Elastica has demonstrable chemical reducing power, as it reacts with certain metallic acids, permanganate and osmium tetroxide (Hass, 1939a, b). This reducing power enables elastica to be demonstrated in tissues by the local reduction of ferric ferricyanide to the insoluble blue ferric ferrocyanide (Cooper, 1969). The autofluorescence and normal optical isotropy of these fibres has been mentioned above.

Elastic fibres and laminae appear in ultrathin sections to be largely composed of a rather pale central amorphous region surrounded by condensations of finely fibrillar electron-dense material (Rhodin, 1962a; Greenlee, Ross and Hartman, 1966; Ross and Bornstein, 1969; Gerrity and Cliff, 1972) (figure 52). Exactly how the surrounding feltwork of fine fibrils is related to the probable underlying fibrillar framework of the elastic tissue (see p. 111) is not at all clear. The diameter of these fibrils is 9–10 nm (Haust et al., 1965; Greenlee et al., 1966; Gerrity and Cliff, 1972) and their general similarity to the fine fibrils of the connective

114 The extracellular components of blood vessel walls

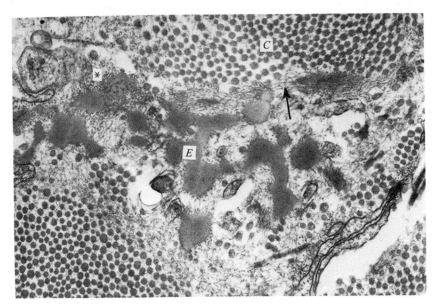

Figure 52. A collection of small elastic fibres (*E*) within the aortic adventitia of a mature rat. There are numerous electron-dense fine fibrils related to the edges of the elastic fibres. They are cut either in cross-section (∗), or longitudinally when they appear to form a branching framework (arrow). *C*, collagen. Magnification × 30000.

tissue space has been stressed (Haust, 1965). In the course of elastogenesis the first recognizable stage in the laying down of elastic elements is the formation of electron-dense condensations of such fibrillar material, which rapidly develop central pale amorphous areas characteristic of elastic tissue (Fahrenbach *et al.*, 1966; Greenlee *et al.*, 1966). Ross and Bornstein (1969) considered that this fibrillar material was biochemically distinct from elastic tissue and that its function was to supply the initial scaffold for the laying down of the elastic tissue. Elastic tissue may at times be seen to have collagen fibrils embedded within it, these having become trapped therein during development (Gerrity and Cliff, 1972).

The role of vascular smooth muscle cells in elastogenesis has been discussed in the section on smooth muscle. The extreme insolubility of elastin would apparently demand that some soluble precursor should be initially secreted by the cells. This putative material has been termed 'tropoelastin' by analogy with collagen (Jackson and Cleary, 1967) but attempts to isolate it during normal elastogenesis have so far failed.

Elastica invariably occurs in association with collagen and, as is the case in blood vessel walls, it also frequently occurs in association with smooth muscle cells, which had led to the concept of a myo-elastic form of tissue structure (Hass, 1939*a*, *b*).

The large arteries that receive the full brunt of each cardiac ejection are extremely rich in elastic tissue. These *elastic arteries* are the pulmonary artery, the aorta, and the larger branches of the aorta such as the brachiocephalic and the carotid arteries in birds and mammals. In fish, amphibians and reptiles the bulbus arteriosus is similarly very rich in elastic tissue (Johansen and Martin, 1965). The elastic tissue within the media of systemic elastic arteries is present as laminae which form a series of concentric fenestrated tubes extending along the lengths of such vessels. This description is oversimplified since the laminae fuse with one another in places to give an appearance of branching when viewed in section. In addition to the elastic laminae there is quite a significant quantity of interlaminar or branch elastica running through the smooth muscle and collagenous tissue that separates the elastic laminae in mammals (Franklin and Haynes, 1927; Keech, 1960a; Pease and Paule, 1960; Paule, 1963; Cliff, 1967, 1970; Wolinsky and Glagov, 1967a). In avian elastic arteries, by contrast, the elastic tissue is confined to fibro-elastic tracts that separate the alternate layers of smooth muscle cells in the vessel wall (Siller, 1962; Moss and Benditt, 1970). The actual amount of elastic tissue present within the walls of elastic arteries has been determined chemically. Hass (1943) found the average content of purified elastic tissue (elastin) within the human aorta to be 38% of its dry weight. Gerrity (1972) found an elastin content of 35% of the dry weight in the thoracic aorta of the adult rat. It is interesting to note that Gerrity found that the volume occupied by elastic tissue, as determined stereologically from electron micrographs, was 52% of the aortic media. Grant (1967) determined the amounts of elastin that could be extracted from various sites along the lengths of porcine, ovine, hircine and human aortae. His results illustrated a longitudinal gradient in the elastin content within elastic arteries, with the highest content proximally, progressively declining in amount as the vessel extended distally. Using his results for porcine aorta as the most graphic example, it can be shown that the elastin content fell from 51% in the arch to only 9% in the lower abdominal aorta. This gradient can be appreciated histologically, as for instance in the giraffe's carotid arteries, which proximally are elastic arteries and distally are muscular-type arteries (Franklin and Haynes, 1927). Rees and Jepson (1970) found that the elastic tissue in the media of carotid arteries of rabbits, cats and dogs occupied from 20–30% of the volume of the media except in the carotid sinus region where it occupied 62%.

The pulmonary artery by contrast, although it contains a great deal of elastic tissue within its media, does not normally possess the laminated elastic structure of systemic elastic arteries (figure 53). It is interesting to note, however, that in congenital types of pulmonary hypertension the foetal aortic configuration of elastic tissue persists within the pulmonary arterial wall (Heath et al., 1959).

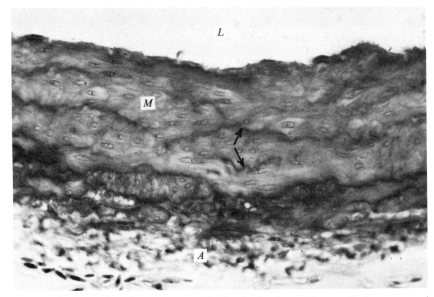

Figure 53. The pulmonary artery wall of a budgerigar. The media is composed mainly of smooth muscle cells (*M*) with branched wavy fibres of elastic tissue running between them in places (arrows). *L*, lumen; *A*, adventitia. Resin section: Richardson stain. Magnification ×480.

There is considerably less elastic tissue present in the muscular arteries that arise as branches from the elastic arteries. The internal elastic lamina is usually a conspicuous feature of such vessels, forming a sharp demarcation between the media and the intima. When identified in vessels that have been fixed in the usual manner the internal elastic lamina follows a highly characteristic wavy course, but in vessels fixed by perfusion at arterial blood pressure this lamina is found to be quite smooth (Short, 1966). The waviness normally seen is an indication of arterial contraction (Phelps and Luft, 1969). The lamina is fenestrated in the same manner as those within the elastic arteries (Parker, 1958; Nyström, 1963). Parker (1958) considered that these fenestrae were of importance in providing nutritive pathways linking the intima and media through the relatively impermeable internal elastic lamina. There are only occasional fine fibrils of elastic tissue identifiable within the media of muscular arteries. At the outer margin of the media there is frequently another laminar condensation of elastic tissue present which sharply demarcates the medio-adventitial border. This structure may be very prominent in certain avian vessels where it can overshadow the internal elastic lamina (figure 54).

Internal elastic laminae show considerable variation from artery to artery. The arterial vessels of the uterus are extremely rich in elastic tissue

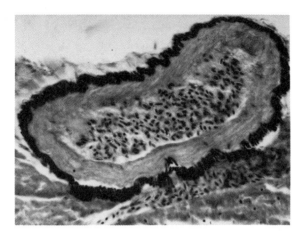

Figure 54. A dense collection of black elastic fibres forms a prominent external elastic lamina in the wall of a coronary artery of a budgerigar. The lumen contains nucleated erythrocytes and no internal elastic lamina can be seen. Verhoeff–Van Gieson stain. Magnification × 300.

(Ramsey, 1955) and the uterine, the mesometrial and the ovarian arteries all show extremely interesting hormone-dependent changes (Hass, 1939a, b). Moritz (1967) described cyclic changes occurring in uterine and ovarian arteries with oestrus and also post-partum in which the media degenerated and the internal elastic lamina became a broad granular band before being displaced peripherally by the formation of a new media and internal elastic lamina. In fact Albert and Bhussry (1967) found that the distribution and number of elastic laminae identified within uterine and mesometrial arteries accurately reflected the number of pregnancies experienced by guinea pigs up to three. Thereafter the pattern could only be identified as multiparous. The human umbilical artery by full term has a well-developed internal elastic lamina which is split into layers by collections of smooth muscle cells (Schallock, 1938). Moore and Schoenberg (1959) found that by the eighteenth week of gestation human umbilical arteries had several interconnected elastic laminae present in their walls.

The arterial limbs of arterio-venous anastomoses (AVAs) have only a fragmented internal elastic lamina which is continuous with the outer elastic network of the venous limb of the structure (Cauna, 1970). Burton (1954) has stated that elastic tissue is not present within the arterial limbs of AVAs at all. Elastic tissue is entirely absent from the walls of retinal arteries (Hogan and Feeney, 1963), which is not, however, a general feature of vessels of the central nervous system (Pease and Molinari, 1960; Nyström, 1963).

An internal elastic lamina can be found in systemic arterial vessels as small as 20 μm in diameter (Short, 1966) and in pulmonary arterial vessels down to 40 μm diameter, which also retain an external elastic lamina down to 30 μm diameter (Alexander and Jensen, 1963).

Franklin (1928) has stated that normally not much elastic tissue is present in the walls of veins. A definite internal elastic lamina has been identified within the rat's portal vein (Ts'ao, Glagov and Kelsey, 1970). In addition Rhodin (1968) has described the presence of elastic membranes within small collecting veins (diameters 100–200 μm) in the muscle fascia of rabbits, whilst Alexander and Jensen (1963) identified an internal elastic lamina within bovine pulmonary veins down to diameters of 50 μm and an external elastic lamina in these veins down to 20 μm diameter. Schallock (1938) considered that the splitting into layers of the internal elastic lamina by collections of smooth muscle cells was a characteristic of veins.

Elastic tissue also occurs as a fibril network interwoven with robust collagen bundles within the inner layers of the adventitia of arteries (Franklin and Haynes, 1927). Fibres of elastic tissue may be identified within the intima of arteries, usually in association with collections of smooth muscle cells. Such myo-elastic cushions, as found just distal to branches of cerebral arteries, were considered to have physiological functions by Hassler (1962). Wright (1963) considered that musculo-elastic cushions related to arterial branches probably represented growing points within the arterial tree since they could not be identified after adolescence. Greatly increased amounts of elastic tissue develop within the intima of the ductus arteriosus during its obliteration. Similarly, a remarkable ability of the aortic intima to produce elastic tissue is found during the intimal repair reaction occurring after necrosis of the media (Taylor, Baldwin and Hass, 1950). In this case the exuberant intimal fibroplasia resulted in the formation of new elastic laminae which, with interposed smooth muscle cell layers being developed, imitated the structure of the defunct media.

The function of elastic tissue within the walls of blood vessels is essentially that of providing a readily deformable, highly compliant honeycomb of material that encircles the vessel lumen. Its property of returning to its original shape and dimensions once the force (or stress) responsible for its distortion (or strain) ceases to act is the reason for the name 'elastic tissue' being conferred upon this material. Burton (1954), quite properly, has drawn attention to the fact that, in terms of physics, high elasticity denotes a great resistance of the material under consideration to deformation by an applied stress. This property is expressed in terms of the Young's modulus (Y) of the material where

$$Y = \frac{\text{force/unit cross-sectional area}}{\text{increase in length/original length}}$$

with dimensions dynes cm^{-2}. In fact elastic tissue has both a low modulus of elasticity (3×10^6 dynes cm^{-2}) and a low tensile strength (1×10^7 dynes cm^{-2}) when compared to collagen (Burton, 1954, 1967; Glagov and Wolinsky, 1963). The important feature that elastic tissue does possess, which distinguishes it from other components of blood vessel walls, is an extremely high ($\sim 300\%$) extensibility (Burton, 1967; Roach, 1970). As pointed out in the previous section the physical properties of blood vessel walls are very largely dependent upon the collagen fibres present therein, whilst the cellular components and elastic tissue act to modify the net properties of the vessel walls. Hence, in the large elastic arteries it has been proposed that the interconnecting framework of elastic tissue has the function of spreading stresses uniformly throughout the collagen framework of the vessel wall (Glagov and Wolinsky, 1963; Wolinsky and Glagov, 1964). These authors considered that the collagen and elastic fibres formed a two-phase system analogous to a material such as fibre-glass, whose properties are not merely a summation of those of its individual components but derive from the mechanical interaction of the matrix and fibrillar components. It is interesting to note that the smooth muscle cells of elastic arteries, which act to modify the active tension within the wall, are attached to the elastic tissue elements (Keech, 1960a, b; Pease and Paule, 1960; Wolinsky and Glagov, 1964; Cliff, 1967, 1970). In this manner, it can be envisaged that a rather smooth and adaptable coupling can be maintained between the strong and unyielding collagen fibres and the soft and easily damaged living cellular components of the wall.

The elastic tissue within the walls of the muscular arteries and veins was considered by Burton (1951, 1954) to have two functions. One was to supply a certain level of maintenance tension, by virtue of which vessels can maintain stable equilibria with varying internal hydrostatic pressures. The other was to act in co-operation with the smooth muscle of the wall to make possible stable graded vasomotor responses. He maintained that in the absence of elastic tissue, as for instance in arterio-venous anastomoses, there could only be an all-or-none and not a graded response.

Elastic tissue develops within both arteries and veins in response to sufficiently high levels of wall tension (Hass, 1939a, b; Burton, 1951, 1954). The simplest expression for wall tension, which excludes the factor of wall thickness, is given by the Laplace equation $T = Pr$, where T is wall tension (dyne cm^{-1}), P is hydrostatic pressure (dyne cm^{-2}) and r is the radius (cm). In the case of veins their low intraluminal pressures are offset by their having diameters considerably greater than arteries. Hass (1939a, b) proposed that rhythmic and fluctuating forces were of especial importance in influencing the formation and development of elastic tissue. This concept is supported by the detection of elastic tissue developing most prominently along lines of tension related to beating cardiac explants *in vitro* (Bloom, 1930), in cardiac scars and in fibrous adhesions and scars in serosae (Bunting, 1939; Williams, 1970). Similarly, the development of elastic tissue within vascular prostheses (Warren and Brock, 1964; Jennings, Brock and Florey, 1966) can be linked to the pulsatile forces acting upon them. However, in the case of a segment of autologous vein being implanted into the abdominal aorta there was no new elastic tissue formed within the dilated and hypertrophied graft (Rivkin, Friedman and Byers, 1963). The importance of a sufficiently high level of wall tension for the development of the aortic elastic laminae *in embryo* has been stressed by Arey (1963) when, in the case of an acardiac twin, this vessel developed simply as a muscular artery. Gerrity (1972) found in the post-natal development of the rat's aorta that there was a seven-fold increase in the level of wall tension in the eight weeks following birth and a five-fold increase by weight in elastica present in this vessel over the same period. The level of wall tension also appears to be of importance in maintaining the elastic tissue within the walls of vessels, hence Gray and others (1953) concluded, from a study of human blood vessels, that arteries operating with high wall tensions showed a progressive increase in elastic tissue with age, whilst vessels, such as the pulmonary artery, with low wall tensions showed a decrease in the amounts of elastic tissue present in their walls with age. Hypertension increases the absolute amount of elastin (and collagen) present within the rat's aorta (Wolinsky, 1970, 1971).

Probably the most interesting correlation between wall tension and the elastic tissue present in the wall is that established by Wolinsky and Glagov (1967a). They examined ten mammalian species, ranging in size from the mouse (28 g) to the sow (200 kg), and found that although wall tensions in the various aortae ranged from 7820 dynes cm^{-1} for the mouse up to 203 000 dynes cm^{-1} for the sow the average tension per lamellar unit remained remarkably constant, being between 1090 dynes cm^{-1} and 3010 dynes cm^{-1}. They suggested that the 'lamellar unit', composed of elastic tissue, smooth muscle and collagen, was the basic structural unit of the aortic wall.

Isotope labelling studies have shown virtually no turnover of elastin in tissues (Partridge, 1962). Tao *et al.* (1962) could detect turnover of elastin in rats of five weeks of age but not at any other age up to two years old, and Walford, Carter and Schneider (1964) found evidence for a gradual dilution of label in growing rats, but no evidence of removal of elastin. It is therefore not surprising to find that the elastin content of vessels generally does not decline with age (Hass, 1943; Kraemer and Miller, 1953; Kao and McGavack, 1959; Lansing, 1959) but may even increase (Gray *et al.*, 1953; Albert and Bhussry, 1967). The impression that the elastic tissue content of the aorta declines with age (Burton, 1951) very probably derives from its altered histological distribution. The elastic laminae in such vessels become frayed and thinner (Dempsey and Lansing, 1954) but *pari passu* an increasing amount of elastic tissue builds up within the interlaminar tracts (Cliff, 1970) (figure 55). In effect there occurs a redistribution of elastic tissue within the aortic wall, as proposed on theoretical grounds by King and Lawton (1950). This appears to be an expression of mechanical wear-and-tear and as such represents a form of mechanical senescence (Comfort, 1964).

Feldman and Glagov (1971) detected an altering distribution of elastic tissue within the aortic media by a biochemical technique. Essentially, the innermost regions of the media in children start off being richest in elastin but with aging the gradient reverses so that in old adults this region becomes the poorest in elastin and the adventitial, or outer, regions of the media are the richest in elastin. The redistribution of elastic tissue leads to changes in the physical strength of the elastic network of the aorta. Hass (1943) found that the purified elastic networks obtained from young aortae had considerably higher tensile strengths than those obtained from older aortae. The elastic laminae of the networks exhibiting low tensile strengths had conspicuous discontinuities.

In addition to these physical changes there also occur very definite chemical alterations in elastic tissue with age, so that increasing amounts of dicarboxylic amino acid residues can be isolated from 'old' elastic tissue (Dempsey and Lansing, 1954; Lansing, 1954; Partridge, 1962). This altered amino acid composition may be due to a build-up of a second protein, rich in polar residues, within old elastic tissue. Associated with this altered chemical composition there is a decrease in the degree of acidophilic staining of elastic tissue with age (Menzies and Roberts, 1963).

Elastic tissue has a well-recognized ability to bind cations such as silver (Hass, 1939a, b) and calcium (Martin *et al.*, 1963) and a second age-associated change is the progressive mineralization of elastic tissue (Dempsey and Lansing, 1954). It is important to note that this mineralization occurs within the elastic and not the collagenous tissues of the arterial wall (Lansing, 1959). Partridge (1962) considered that calcification occurred within the fine structure of the elastic fibre and caused increased rigidity of these structures. Such changes must of necessity affect the overall properties of the arterial wall as a two-phase system (p. 119).

Some appreciation of the extent of the mineralization that occurs can be obtained from the results of Lansing (1954) who found a calcium content of 1.14% by weight in elastic tissue of young arteries and 6.39% in that of old arteries. Gray and others (1953), using a micro-incineration technique, established that the severity of calcific changes within elastic tissue increased from pulmonary artery (least severe) to aortic arch to descending aorta and finally to the abdominal aorta (most severe). They also linked the severity of the changes with the presence or absence of

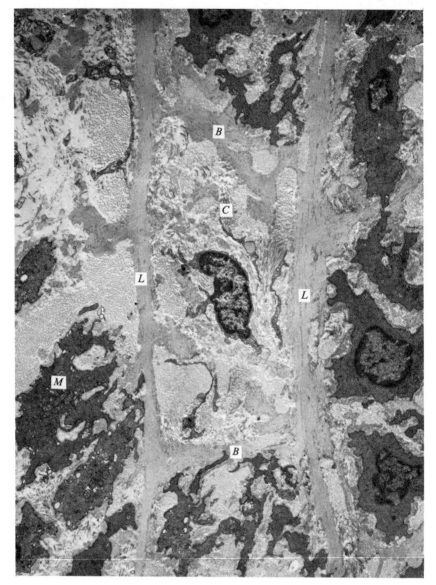

Figure 55. The aortic media of a 3-year-old (senile) rat. The elastic laminae (*L*) are thin and much of the elastica of the media is present in large branches (*B*) arising from the laminae. Note the plentiful collagen (*C*) and loss of cellularity of the media. The remaining muscle cells (*M*) are hypertrophied. Magnification ×6000.

hypertension (higher incidence in negroes than whites) but not with any increased incidence of coronary atherosclerosis or of myocardial infarcts (higher incidence in whites than negroes). The internal elastic laminae of the large arteries of the leg similarly show progressive calcification with age but Wright (1963) found no relationship between the severity of this process and the presence of hypertension or atheroma.

ATHEROSCLEROSIS

In atherosclerosis one of the earliest changes noted histologically is fraying and reduplication of the internal elastic lamina of the affected vessel. The importance of changes in the elastic tissue of arteries in the possible aetiology of atherosclerosis has been stressed (Lansing, 1954, 1961). Gillman (1959) considered that disordered mucopolysaccharide turnover associated with growth and wear-and-tear of vascular elastic laminae might well be the factor initiating accumulation of lipids and fibrosis in atheroma. The possibility that enzymes of the elastase type might be involved in the induction of atherosclerotic changes has been investigated, but with negative results (Tennent *et al.*, 1956; Hall, 1961). Loeven (1969) found small but consistent quantities of elastoproteinase and of elastomucase within bovine aortae; however those fractions which contained such enzyme activity also contained elastoproteinase inhibitor. The breakdown and removal of elastic tissue in arterial walls is accomplished by a foreign-body type of tissue reaction (Gillman *et al.*, 1957). In the absence of such a reaction the elastic membranes of devitalized arterial walls persist indefinitely, showing splitting and a progressive calcification (Zellweger, Chapuis and Mirkovitch, 1970).

ANEURYSMS

Arterial aneurysms occur essentially as a result of changes in the elastic tissue components. The change may be extremely clear cut, as in the case of berry aneurysms of the intracranial arteries related to the Circle of Willis, where the loss of elastic tissue from the arterial wall is the *sine qua non* for their development (Roach, 1970). Nyström (1963) studied the walls of intracranial aneurysms electron microscopically and found hypertrophy and splitting of the elastic laminae with complete loss of elastic tissue near sites of actual rupture of the wall. The elastic tissue appeared very granular and contained electron-dense particles 50 nm to 0.5 μm in diameter.

Aneurysms of the aorta and its larger branches are also associated with loss of elastic tissue. In the abdominal aorta such lesions are linked with atherosclerosis, whilst in the thoracic aorta aneurysms usually develop secondary to syphilitic aortitis. The chronic inflammatory infiltrate around the vasa vasorum that occurs in the tertiary form of this disease is extremely deleterious to the elastic tissue, as is the case of inflammatory changes in general (Hass, 1939a, b). In Marfan's syndrome, which may lead to spontaneous rupture of the aorta, there is found a scattering and disappearance of elastic tissue in the vessel wall (Voigt and Hansen, 1970). Certain animals when fed a copper deficient diet also show a high incidence of death due to aortic rupture (Hill, 1969). The aortae of such animals show severe degenerative changes of elastic tissue within their walls (Simpson and Harms, 1964). Hill (1969) considered that there was improper formation of elastic tissue in such animals due to decreased synthesis of desmosine cross-linkages. He considered, from knowledge of other known amine oxidases, that an essential step involving the oxidative deamination of lysine would require a copper-containing enzyme.

Such cases associating decreased amounts and altered morphology of elastic tissue within the arterial wall with the development of aneurysms and eventual rupture of the wall has led a number of people to equate the elastic tissue directly with the mechanical strength of the wall. In fact from a consideration of its physical properties (p. 118) such a proposition is not tenable. It is most probable that the defects that occur in the elastic framework of the wall result in intolerable local stresses being placed upon the collagenous elements of the wall, due to a failure in the correct functioning of the two-phase system (p. 119). This concept has been used by Roach (1970) to explain the formation of berry aneurysms. As an analogy I would compare an Indian fakir lying on a bed of nails, showing diffusely spread stress, to the same man being left with only, say, half a dozen upturned nails upon which to rest – the condition of poorly diffused stress. It is this failure of the elastic tissue to efficiently spread stresses throughout the arterial wall that results in irreversible distension of the collagenous framework which leads to progressive dilatation and ultimately to rupture of the vessel.

6

ANCILLARY STRUCTURES OF BLOOD VESSEL WALLS

VASA VASORUM

Just as big fleas have little fleas, so big vessels have little vessels within their walls (figure 56). These are the vasa vasorum which show greatest development within those vessels that conduct blood with low levels of oxygen tension. Hence, the walls of the pulmonary artery and the systemic veins of human beings have rich vascular supplies which are completely out of proportion to those normally found in the arterial walls (Winternitz, Thomas and Le Compte, 1938). Their function is to provide nutrient vessels that subserve the metabolic requirements of the vessel wall when its thickness exceeds a certain critical magnitude. This critical thickness, that renders diffusion from the vessel's own lumen insufficient, has been estimated for human aortic tissue by Kirk and Laursen (1955). They calculated that the depth to which oxygen could penetrate the aortic wall by diffusion alone was 0.9 mm in young adults and 1.00 mm in middle-aged adults. They computed that the thickness of avascular media in the human aorta was 0.78–0.99 mm in young adults and 0.97–1.22 mm in middle-aged adults. Hence the statement made by Whereat (1967) that the inner arterial wall is constantly on the verge of anoxia is not over-dramatic. Wolinsky and Glagov (1967b) found a most interesting expression of this critical thickness in investigating aortae from twelve mammalian species, ranging in size from the mouse to the bull. They found that in aortae that had no more than 29 elastic laminae there were no vasa vasorum within the media. In aortae that had more than 29 elastic laminae vasa vasorum were present within the media, but there always remained an avascular subintimal zone of the media which was generally 29 laminae thick. The zone in which vasa vasorum were present was the region of the media in which growth processes, associated with the normal development of the animal, could be identified in the aorta.

The nutrition of blood vessel walls relies on blood–tissue exchange occurring via both the endothelium lining its own lumen and also the endothelium of its vasa vasorum (Higginbotham, Higginbotham and Williams, 1963). The existence of two diffusion gradients extending from the inner and outer regions of the wall into the avascular zone of the media has led to the concept of the 'watershed' effect whereby the region

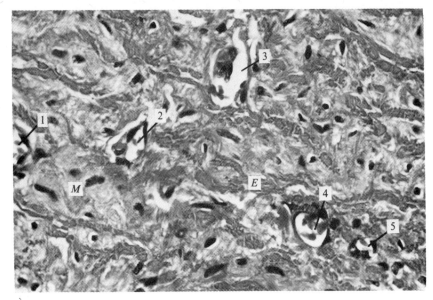

Figure 56. Vasa vasorum (numbered 1–5) penetrating to the junction of the middle and inner thirds of the pig's aortic media. *M*, smooth muscle; *E*, elastic laminae. Photomicrograph, H. and E. stain. Magnification × 530.

situated midway between these two sources is living more or less on the border line of viability. Hence, it is the mid and inner zones of the aortic media that show progressive loss of enzyme activity in aging human beings, being termed 'anoxic enzyme failure' by Adams, Bayliss and Ibrahim (1963). Similarly, Garbarsch *et al.* (1969) found that the most pronounced aortic lesions produced by repeated bouts of hypoxia in rabbits were present within the luminal half of the media. A very striking selective localization of medial degeneration has been obtained in the middle third of the aortic media in uraemic rats (Carter *et al.*, 1966).

Zellweger, Chapuis and Mirkovitch (1970) completely isolated vasa vasorum in segments of the dog's thoracic and abdominal aorta from any blood supply by separating off the adventitia, tying and cutting the related intercostal and lumbar arteries and then enclosing these segments of aorta within snugly fitting sleeves of silicone rubber. From one to six days later necrosis of the outer zones of the aortic media occurred but there remained a very clear-cut subintimal zone of the media 0.48 mm in thickness where the medial smooth muscle cells survived (figure 57). These experiments illustrate two things. Firstly, the importance to the vessel wall of the blood supply provided by the vasa vasorum and, secondly, the very definite role that diffusion from the vessel's own lumen has in maintaining the viability of the subintimal zones of the media. The possibility of damage occurring to the vasa vasorum when a vessel is

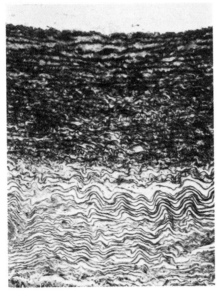

Figure 57. The result of depriving the dog's aortic media of its vasa vasorum. The luminal third (upper half of figure) is cellular and viable whilst the middle third (lower half of figure) and outer third are acellular and necrotic. Photomicrograph taken 10 days after wrapping aorta. Magnification × 100. Reproduced through the kindness of Dr J. P. Zellweger.

mobilized in the course of surgery must be borne in mind and Smith (1967) has strongly recommended that care be taken to retain the adventitia in the course of the microvascular suturing of blood vessels.

The distribution and origins of the vasa vasorum of a variety of blood vessels have been studied in a series of animals, generally either by the technique of filling the vessels under pressure with some readily visible injection mass and subsequently clearing the tissues (Woodruff, 1926; Winternitz et al., 1938) or else by injecting the vasa vasorum with a radio-opaque material and examining the vessel walls by an X-ray microscopic technique (Schlichter and Harris, 1949; Clarke, 1965).

Vasa vasorum most commonly originate at the first small branches of tributaries that arise from the larger vessels, such as the coronary, intercostal and lumbar arteries in the case of the aorta (Woodruff, 1926; Robertson, 1929; Winternitz et al., 1938; Woerner, 1951; Clarke, 1965). Generally the small arterial vasa arise from branches almost immediately they are clear of the adventitia of the parent vessel. Robertson (1929) studied the vascularization of the thoracic aorta in dogs, lambs and human beings and concluded that the general pattern was that vasa arose from branches about 2 mm outside the aortic adventitia and ran back to supply the aortic wall with numerous anastomoses with one another. Winternitz and others (1938) studied aortic vasa vasorum in human beings and a variety of other mammals and found that they could even arise at times within the lips of large branches where they arose from the aorta. In certain mammals, such as the dog and the cow vasa can also arise directly from the aortic endothelial surface to nourish the underlying media. Finally, some vasa may arise from the blood vessels of the peri-aortic fat.

By a process of branching and anastomosis the vasa vasorum of the aorta form a network of vessels which is richest in relation to the origins of large branches (Robertson, 1929). From the network of vessels in the adventitia, branches penetrate the aortic wall to supply the media (Woodruff, 1926; Winternitz et al., 1938). Two main vascular networks, one at the medial–adventitial border and the other at the junction of the outer and middle thirds of the media have been identified in the canine aorta (Woerner, 1951). The dog, however, is distinguished by the extremely rich vascularity of its aorta compared to that of the human being, the chicken and the rabbit (Schlichter and Harris, 1949). The importance of species differences in the richness of aortic vasa vasorum is emphasized also by the findings of Winternitz and others (1938) who could readily identify rich networks of vessels within the aorta of the cow and the buffalo but much sparser ones in undiseased human aortae.

Vasa vasorum are usually limited to supplying the outer two-thirds of the aortic media (figure 56), but in the equine aorta vasa vasorum arising from the adventitia normally ramify along the lines of the smooth muscle fibres to reach the border of the intima (Woodruff, 1926). Very rarely capillaries may be identified within the inner third of the canine aorta (Woerner, 1951). Clarke (1965) in a most valuable study of the development of the vascularization of the human aorta traced the progressive ingrowth of vasa vasorum into deeper and deeper layers of the media. At four years of age arterial vasa were present in the outer third of the media, at ten years of age in the middle third and by thirteen years of age a capillary and venule bed was present in the inner third of the media. In no instances have vasa vasorum been identified within the undiseased intima (Geiringer, 1951; Woerner, 1951).

The venous drainage of the vasa vasorum of the aortic media is into the adventitial venous plexus and thence into the cardiac, bronchial, intercostal and lumbar veins (Clarke, 1965).

The venous drainage of the vasa in the walls of veins can occur via two possible routes. Franklin (1928) considered that they drained directly into the lumen of the parent vessel by way of small collecting veins which penetrated the wall. On the other hand Higginbotham and others (1963) considered that the general view was that the vasa of veins anastomosed in the adventitia to form collecting veins that drained into adjacent veins.

The vasa vasorum are specialized components of the microcirculation whose role is intimately linked to the physiological functioning of the blood vessel wall which they supply (Higginbotham et al., 1963). It is therefore surprising that so little is apparently known about their normal physiology. Duff (1932) demonstrated their permeability to Trypan blue and showed that the patchy staining produced in the aorta was in fact related to the varying density of the vasa vasorum present in the wall. However, studies on the passage of lipoproteins through the aortic wall have revealed no role being played by these vessels (Watts, 1961; Adams et al., 1964). The haemodynamics pertaining to these specialized microcirculatory beds in the arterial walls are most unusual. In the first place these microcirculatory systems arise almost directly from the large arterial lumen with only a short intervening segment of narrow arterial vasa present. This means that there can be very little 'smoothing' of the pulse wave, resulting in considerable pulse pressure and intermittent flow within these microcirculatory beds. Secondly, it is hard to visualize exactly how the flow within and the permeability of these vessels will be related to the changes occurring in intramural tension within the artery they supply. Most probably the situation is analogous to that which occurs in the coronary circulation where flow within these vessels virtually ceases during cardiac systole (Gregg, 1934).

VASA VASORUM AND DISEASE PROCESSES

The presence of microcirculation within the walls of blood vessels sets the stage for inflammatory, exudative and ischaemic lesions to be produced therein. Winternitz and Le Compte (1940) studied infectious angiitis induced experimentally with various bacteria and found that the walls of arteries were very resistant to infection, the only lesions being produced by *Staphylococcus aureus*. By contrast acute suppurative lesions were readily produced in the walls of veins. They found that connections between the vasa of related arteries and veins permitted the spread of infection from one to the other. Robertson (1929) found that vasa were most numerous in those sites of the thoracic aorta that were readily attacked by toxic and infectious conditions and were sparsest in those sites where senile changes developed most readily. The higher incidence of atherosclerosis in the abdominal aorta as compared to the thoracic aorta may therefore be related in some way to the richer vascularity of the former (Clarke, 1965). A similar correlation between richness in vascularity of the arterial wall and incidence of atherosclerotic lesions was found by Winternitz and others (1938). Whether this correlation indicates a primary involvement of the vasa vasorum in the disease process, as for example by providing the pathway for local exudation of plasma and lipids as well as cells into the wall, or a secondary relationship in so far as their presence might indicate that that particular region of the wall is more metabolically vulnerable to attack by disease processes, is not clear.

From the point of view of this latter proposition the experiments of Schlichter, Katz and Meyer (1949) are informative. They found that the normally high refactory state of the canine aorta to the experimental induction of atherosclerosis could be overcome in areas where the extremely rich vasa vasorum of this species had been destroyed. The normal arterial intima contains no vasa vasorum but if its thickness comes to exceed a certain critical value, which in the case of the aorta is 0.5 mm, then minute vessels grow into the intima, from the lumen or the media, or both (Geiringer, 1951). Geiringer considered that new vasa arose directly from the lumen in the course of organization of mural thrombi and that the luminal and medial sets of vessels invading the intima could anastomose with each other. The capillary vessels within the atherosclerotic intima are similar in fine structure to capillaries that occur elsewhere in the body (Buck, 1959) and once a microcirculatory system has developed therein then true exudatory and inflammatory reactions, haemorrhage and infarction can occur within the intima (Le Compte, 1967). Winternitz and others (1938) stressed the frequent occurrence of haemorrhages from the small vessels within atheromatous lesions. They considered that such haemorrhages and acute exudative processes occurring within the thickened intima could act either to precipitate thrombosis of the artery or to directly close the arterial lumen by suddenly increasing the size of the plaque. Higginbotham and others (1963) placed most importance on necrosis supervening within atherosclerotic plaques due to their failure to acquire or to retain an adequate blood supply and concluded that the vast majority of vascular accidents associated with atherosclerosis could be attributed to this cause.

LYMPHATICS OF BLOOD VESSEL WALLS

Lymphatic vessels are present within the walls of both arteries and veins (Higginbotham *et al.*, 1963; Johnson, 1969; Yoffey and Courtice, 1970). In the case of arteries lymphatic vessels are confined to the adventitia and do not enter the media or intima (Watts, 1961; Johnson and Blake, 1965; Dunihue, 1967; Jellinek *et al.*, 1970). In veins, on the other hand, these structures may penetrate more deeply into the vessel wall, at times coming to lie within the media. When rendered visible by injection, lymphatics are seen to form fine networks within the vessel walls and they may be traced into continuity with the general lymphatic system. Hence, those of the aorta can be seen to drain into the thoracic duct either directly or via regional lymph nodes whilst those of the coronary arteries are seen to drain into the subepicardial lymphatic plexus of the heart. It is interesting to note that Johnson (1969) identified small lymph nodes actually within the adventitia of the human aorta. The lymphatics of blood vessel walls conform in their appearance to lymphatic vessels elsewhere in the body

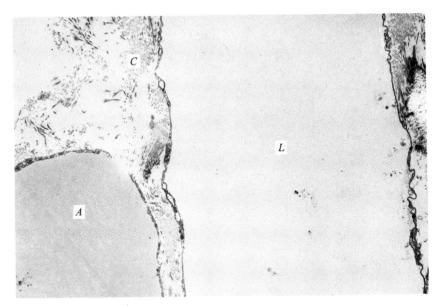

Figure 58. A lymphatic vessel situated in the outermost adventitia of the rat's aorta. The wide lumen (L) is lined by a tenuous endothelium which is intimately related to the collagenous connective tissue (C). A, adipose cell. Magnification ×4000.

(Yoffey and Courtice, 1970). This similarity is confirmed also at the ultrastructural level in electron microscopic studies of these structures (Jellinek *et al.*, 1970) (figure 58).

It is presumed that their function within blood vessel walls is identical to that of lymphatic vessels elsewhere, which is to drain off water and contained solutes, particularly macromolecules and lipids, which would otherwise accumulate within the tissues following their passage through the blood vascular endothelial membranes (Yoffey and Courtice, 1970). In the case of arterial walls this occurs both through the endothelial lining of the artery itself and also through that of the vasa vasorum of the vessel (Cowdry, 1952; Higginbotham *et al.*, 1963). It is thought that there is a progressive 'milking' of fluid through the arterial wall from the intima to the adventitia, associated with the pulsatile pressure changes occurring rhythmically within the wall (Cowdry, 1952; Taylor, 1955; Higginbotham *et al.*, 1963; Johnson, 1969). In the case of the aorta the very large endothelial surface in comparison with the actual mass of aortic tissue has been remarked by Vost (1969), and permeability of its intima for protein (Packham *et al.*, 1967) and cholesterol (Adams *et al.*, 1964) has been demonstrated. In the case of muscular arteries, such as the human coronaries, outward passage from intima to adventitia of serum β-lipoprotein, albumin and γ-globulin has been demonstrated (Watts,

132 Ancillary structures of blood vessel walls

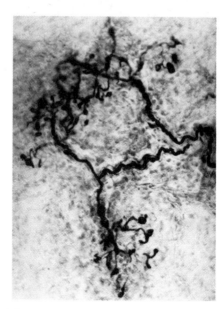

Figure 59. Trefoil and fan-shaped baroreceptor end organs in the aortic arch of a pig. (Abraham, 1969.)

1961). Whilst an important role has been deduced for the adventitial lymphatics in such vessels, experimental observations on their function in the uptake of materials traversing the arterial wall are few. Watts (1961) demonstrated the entry of normal serum protein constituents into the adventitial lymphatics, and Jellinek and others (1970) demonstrated the entry of lipids into such vessels related to the aorta of hypercholesterolaemic rats.

LYMPHATICS AND DISEASE PROCESSES

Robertson (1929), in reviewing earlier investigations of the aortic wall, commented on the possible spread of syphilis via the lymphatic vessels of the aorta. Interference with the normal removal of lipids by the arterial lymphatics has been proposed as being of pathological significance in the development of atherosclerosis (Taylor, 1955; Jellinek et al., 1970). Johnson (1969) has drawn attention to the fact that ingrowth of vasa vasorum occurs in response to increased metabolic demands within arterial walls but that ingrowth of lymphatics does not occur in these regions. In this way the stage is set for increased passage of materials into the arterial wall with no concomitant route of escape being established by lymphatic ingrowth. This could certainly be a possible cause of lipid accumulation within intimal plaques which have acquired vasa vasorum.

INNERVATION OF BLOOD VESSEL WALLS

Innervation of blood vessels has been studied by a variety of techniques, each of which is capable of yielding information that is unobtainable by the other techniques. Hence the methods currently used are complementary to one another and no one technique can be regarded as superseding any other. The property of living nerve fibres to stain preferentially with methylene blue is of great value in identifying these structures *in vivo* and in tissue spreads (Clark, Clark and Williams, 1934; Clark and Clark, 1947; Abraham, 1969). Extremely beautiful and detailed demonstrations of the axons and nerve endings within blood vessel walls may be obtained with the technique of silver impregnation, often with post-treatment with gold, as seen in the work of Abraham (1969) (figure 59). In view of the fact that 'making visible the vascular nervous system is still one of the most difficult tasks of neurohistology' (Abraham, 1969), the development of the Falck–Hillarp technique (Falck, 1962, 1964) which produces autofluorescence of catecholamine-containing nervous tissue (figure 60) has acted as a great stimulus to the study of blood vessel adrenergic innervation. The histochemical demonstration of acetylcholinesterase is of value in demonstrating cholinergic elements within vessel walls (Abraham, 1969; Fillenz, 1970; Silva and Ikeda, 1971) and Abraham (1969) found that this technique outlined the shape and descent of aortic nerve fibres remarkably accurately when compared to the results of the silver impregnation technique.

The application of electron microscopy to identify autonomic nerve endings related to blood vessel walls has led to a new conception of the relationship between the elements of the neuromuscular synapses. The sites of neurotransmitter release are characterized by localized increase in width of the axons (Simpson and Devine, 1966) which are deficient in Schwann cell and basement membrane covering in these regions and contain numerous large mitochondrial profiles (Burnstock and Merrillees, 1964; Lever *et al.*, 1965; Lever, Spriggs and Graham, 1968; Verity and Bevan, 1968) and cytoplasmic vesicles. It is these vesicles that are involved in the storage and release of the neurotransmitter substances that act upon the blood vessel wall (Burnstock and Merrillees, 1964; Lever *et al.*, 1965; Simpson and Devine, 1966; Lever *et al.*, 1968; Verity and Bevan, 1968; Siggins and Bloom, 1970; Burn, 1971). The presence of extremely electron-dense granules within the vesicles indicates the presence of norepinephrine (noradrenaline) (Burnstock and Merrillees, 1964; Bodian, 1970; Burn, 1971) which can readily be distinguished from palely-staining epinephrine granules in glutaraldehyde-fixed tissues (Coupland and Hopwood, 1966). There are two populations of dense-cored vesicles within the vasomotor nerves, one being a group of small vesicles, with diameters of about 30–50 nm, and the other being a group of large vesicles,

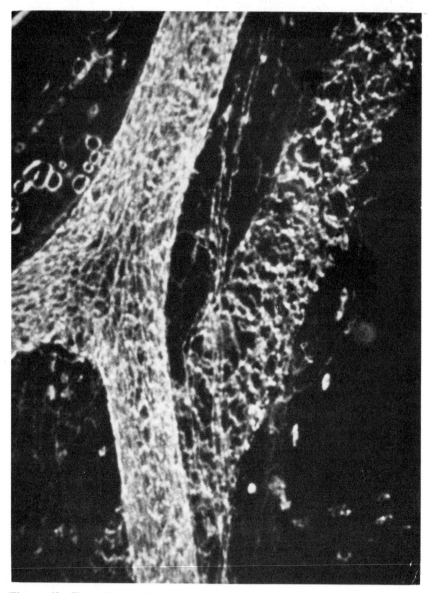

Figure 60. Formalin gas-induced catecholamine fluorescence demonstrated in nerve plexuses related to an artery (dense network) and a vein (open network) of the rat mesentery. (Falck, 1962.)

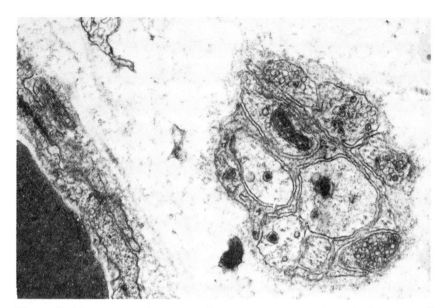

Figure 61. Nine autonomic nerve endings partially enclosed within Schwann cell cytoplasm and approximately 1.0 μm from the wall of a small blood vessel at the lower left of figure. They contain both large and small granular vesicles and small non-granular vesicles. Cytoplasmic microtubules and a single mitochondrion can be identified within the endings. Rat, aortic vasa vasorum. Magnification × 36000.

with diameters about 60–150 nm (Verity and Bevan, 1968; Fillenz, 1970; Burnstock, 1972) (figure 61). The large granular vesicles make up only 3–4% of those present within autonomic nerve endings (Burnstock, 1972). Autonomic nerve endings containing virtually all non-granular vesicles are generally taken as being cholinergic (Bodian, 1970; Burnstock, 1972). The detection of such endings in the vasomotor innervation of the renal medulla led Moffat (1967) to conclude that it was probably cholinergic in type. However, caution in the interpretation of agranular vesicles is indicated by the work of Lever and others (1968) who established the adrenergic character of vasomotor nerves in the pancreas by the fluorescence technique, [^{3}H]noradrenaline uptake by axons and detection of degeneration following sympathectomy, even though the autonomic vesicles were very largely agranular in their micrographs. In addition, Fillenz (1970) found that the dense granules within small autonomic vesicles in pulmonary vessels required glutaraldehyde-fixation for their demonstration.

In contrast to the close associations (20 nm separation) that exist in places between nerve endings and smooth muscle cells in other sites (Merrillees, Burnstock and Holman, 1963; Thaemert, 1963) their relationships in vessel walls are more remote, having been aptly described as '*en passage*' synapses (Burnstock and Merrillees, 1964). In fact, when, as is frequently the case, this relationship occurs across the adventitial–medial border, the distance separating nerve endings from the vascular smooth muscle can be measured in microns. In large arteries this separation may be 2 μm (Silva and Ikeda, 1971) or even as great as 4 μm (Fillenz, 1970). Similarly, in arterioles the separation can be 5 μm and in terminal arterioles 1 μm (Rhodin, 1967). On rare occasions closer association between nerve endings and smooth muscle can be identified in blood vessel walls where minimal separation is between 80 and 100 nm, in round figures (Thaemert, 1963; Lever *et al.*, 1965; Simpson and Devine, 1966; Fillenz, 1970). Separations of less than 80 nm are extremely rare (Burnstock and Merrillees, 1964). Relations as close as 7.5 nm have, however, been identified between nerves and smooth muscle cells in the great vessels of the foetal lamb (Silva and Ikeda, 1971) but such findings are the exceptions rather than the rule. In the cephalic aorta of *Octopus vulgaris* neuromuscular synapses with 10 nm separation of the related membranes and regions of tight junction occur (Barber and Graziadei, 1967). The claim by Burkel (1970) that actual neuromuscular membrane contacts occur in the pre-capillary sphincters of the rat's hepatic arterial system is therefore worthy of further investigation.

DISTRIBUTION OF NERVES IN VESSEL WALLS

In general, vessels that contain smooth muscle in their walls are supplied with non-myelinated nerves of the sympathetic nervous system (Franklin, 1928; Ehinger, Falck and Sporrong, 1966; Abraham, 1969; Burnstock, Gannon and Iwayama, 1970). However, there are certain vessels, such as the spiral arterioles of the endometrium (Ancla and de Brux, 1964), the retinal (Hogan and Feeney, 1963) and the umbilical vessels (Somlyo, Woo and Somlyo, 1965; Bell, 1972) which are non-innervated. Similarly, the blood vessels within the central nervous system are described as almost completely lacking sympathetic innervation (Burnstock *et al.*, 1970). The small exchange, or capillary, vessels are not innervated. Ehinger and others (1966) have shown that adrenergic nerves are not present in systemic capillaries whilst Fillenz (1970) has shown the absence of adrenergic and cholinergic nerves in pulmonary capillaries. Abraham (1969) identified 'accompanying fibres' but could not demonstrate definite nerve endings within the walls of capillaries. Electron microscopy has revealed the presence of autonomic nerves closely related to the walls of pancreatic capillaries, some of which may be innervating the vessels rather than the pancreatic tissue they nourish (Lever *et al.*, 1968). A possible example of motor innervation of capillary endothelium in the terminal hepatic arterial system has been suggested by Burkel (1970) but no definite proof has been obtained for a physiological relationship between nerve endings and endothelial cells. It would appear that the claim by Krogh (1959) that capillaries in general are invested by rich nets of nerve fibres responsible for 'capillariomotor reflexes' is not tenable in the light of present work.

As a general rule the walls of arteries are more richly innervated than those of veins (Falck, 1962; Ehinger *et al.*, 1966; Burnstock *et al.*, 1970; Furness and Moore, 1970). This is most evident in fluorescent preparations which demonstrate thick plexuses of adrenergic fibres surrounding arteries while more open meshworks of fibres surround veins (figure 60). There are certain veins, however, that possess particularly rich innervation. These are the vena saphena magna, which has a very heavy muscular coat, the vena cava (Abraham, 1969), the mesenteric and renal veins (Abraham, 1969; Bennett, 1971), and the renal portal system of birds (Bennett and Malmfors, 1970).

The considerable variation in degree of innervation at various sites within the arterial tree has been stressed (Abraham, 1969). Silva and Ikeda (1971) found that, in foetal lambs, the richness of innervation of the pulmonary artery was greater than that of the *ductus arteriosus* which in turn was greater than that of the aorta. The pulmonary arteries in the lizard are very densely innervated by comparison to those of mammals (Furness and Moore, 1970). Innervation can be traced to the level of the finest arterioles and pre-capillary sphincters (Ehinger *et al.*, 1966; Rhodin, 1967; Lever *et al.*, 1968) where it is generally much less dense than in the arteries (Falck, 1962). However, both limbs of arterio-venous anastomoses are well supplied with nerve fibrils (Abraham, 1969). Innervation reappears in the fine post-capillary veins (Ehinger *et al.*, 1966) which, from the very careful electron microscopic study of Rhodin (1968), correspond to small collecting veins with luminal diameters of 100–200 μm and have three or more layers of smooth muscle cells in their walls. Smaller veins, with diameters of 50 μm and below have been shown to be non-innervated (Lever *et al.*, 1968).

Nerve axons form two interconnecting plexuses: one in the mid, the other in the deep layer of the adventitia of arteries. In general, nerves do not enter the media of either arteries (Burnstock and Merrillees, 1964; Ehinger *et al.*, 1966; Abraham, 1969; Bennett and Malmfors, 1970; Burnstock *et al.*, 1970; Rolewicz, Gisslen and Zimmerman, 1970) or arterioles (Lever *et al.*, 1965; Simpson and Devine, 1966; Rhodin, 1967). This contrasts with the situation in veins where nervous penetration of the media commonly occurs (Ehinger *et al.*, 1966; Rolewicz *et al.*, 1970; Ts'ao, Glagov and Kelsey, 1970). Birds have considerably richer innervation of their veins than do mammals (Bennett and Malmfors, 1970) and in fact Abraham (1969) has drawn attention to the avian pulmonary veins as being the best sites for studying vascular smooth muscle and nerve interrelationships. Occasionally, however, nervous penetration of the media does occur in arteries. Abraham singled out the coronary vessels as being more richly innervated than any other vessels, in particular with relation to the innervation of the media. He also found nerve plexuses whose end fibres extended between the muscle cells of the media of the aorta of the cat, the dog, the fox and the badger, as well as of the arteries supplying the base of the brain and of the splenic and renal arteries in certain animals. Sympathetic innervation has been detected in the outer third of the media of the renal and carotid arteries of the sheep (Burnstock *et al.*, 1970). Rarely, penetration of the media by nerves also occurs within arterioles, as described in the renal cortex of the sheep (Simpson and Devine, 1966) and in the muscle fascia of the rabbit (Rhodin, 1967). Innervation of the outer media of the pulmonary arterial vessels has been described in the dog (Fillenz, 1970), the cat (Verity and Bevan, 1968) and the rabbit (Verity and Bevan, 1968; Burnstock *et al.*, 1970).

In addition to autonomic endings, ganglion cells have been identified within the adventitia of the aortic arch and the coronary arteries and veins in the cat (Abraham, 1969). Thicker myelinated fibres supply sensory baroreceptors within the adventitia of the aortic arch, the carotid sinus, the pulmonary artery and the renal artery. These sensory endings show extremely delicate patterns with trefoil, fan-shaped and ivy-leaf patterns (figure 59). Electron microscopically such sensory endings are found to contain large numbers of mitochondrial profiles and few vesicles (Burnstock, 1972). There is evidence for a special blood supply to the adventitia in the regions of these sensory endings (Abraham, 1969). Chemoreceptors occur in association with blood vessel walls but are not basically part of their intrinsic structure. The excellent reviews of Comroe (1964) and Korner (1971) give a most useful account of these structures and the reader is referred to these for further information. No nerves are found in the intima of vessels (Abraham, 1969).

FUNCTION OF NERVES WITHIN BLOOD VESSEL WALLS
General

In view of the extreme rarity, or even complete lack, of true neuromuscular synapses within blood vessel walls (p. 135) their mode of nervous control is unusual. It is generally envisaged that it is achieved by means of chemical transmitter release from nerve endings located mainly at the adventitial–medial border, which then reaches receptor sites on the vascular smooth muscle cells by diffusion (Burnstock and Merrillees, 1964; Burnstock *et al.*, 1970; Ljung, 1970). This implies that only the outer layers of smooth muscle cells are directly affected by transmitters whilst the remaining cells throughout the wall are activated by electronic coupling. There is also reason to believe that the innermost layers of smooth muscle cells may respond to altering levels of catecholamines in the bloodstream which reach them by diffusion from the vessel lumen

(Mellander and Johansson, 1968; Burnstock et al., 1970; Guntheroth and Chakmakjian, 1971). However, since the level of catecholamines in the bloodstream is measured in nanograms per ml and the levels produced locally at nerve endings are measured in micrograms per ml (Mellander and Johansson, 1968), with the estimated level attained at the smooth muscle receptors being about one fifth of this (Ljung, 1970), then any nerve-mediated effects will completely dominate the vascular reactions.

Sympathetic system

Sympathetic stimulation produces vasoconstriction in most, but not all, arteries and veins. The sympathetic vasoconstrictor control of blood flow is very important in maintaining vasomotor reflex control of the circulation and general cardiac homeostasis (Mellander and Johansson, 1968). The sympathetic vasoconstrictor response in the gut differs from that in skin and muscle in showing an initial short constrictor response to stimulation followed by the phenomenon of 'autoregulatory escape' in the resistance vessels. Sympathetic vasoconstrictor responses are mediated by norepinephrine release from the terminal axons (Burnstock, 1972) and rely for their effect upon the sensitivity of the vascular smooth muscle to α- (excitatory) stimulation (Somlyo and Somlyo, 1970).

The sympathetic nervous system can also bring about vasodilation in certain vessels. As pointed out by Burnstock (1972) this action of the sympathetic system is more complex than its vasoconstrictor role. In the smooth muscle of the coronary vessels of many species this response is probably due to the preponderance of β- (inhibitory) over the α-receptors (Somlyo and Somlyo, 1970). In skeletal muscle sympathetic vasodilatation is cholinergically mediated and forms part of the 'fight or flight' alerting mechanism of the body (Mellander and Johansson, 1968). There is evidence to suggest that this cholinergic form of vasodilatation is under cerebral control (Ellison and Zanchetti, 1971).

An avid catecholamine storage by sympathetic nerves in the walls of veins has been demonstrated (Rolewicz et al., 1970). This may have an important physiological function in preventing spillover of catecholamines from the regional vascular beds into the general circulation.

Parasympathetic system

From in-vivo studies on the reactions of small blood vessels claims have been made for the existence of dual innervation (Lutz and Fulton, 1958). Abraham (1969) identified both ganglion cells and parasympathetic fibres within the extremely rich perivascular nerve plexuses of the coronary arteries. From this point of view it is interesting to note that acetylcholine produces vasoconstriction in porcine coronary arteries, with this effect being reversed by atropine (Somlyo and Somlyo, 1970). Bell (1972), in reviewing autonomic nervous control of reproduction, concluded that the pelvic parasympathetic nerves were responsible for the vasomotor changes necessary to produce penile erection in mammals.

Sensory innervation

Baroreceptor endings, which are forms of stretch receptors, are present in the carotid sinus and the aortic arch and also in the wall of the pulmonary artery main branches. The functions of these structures in the neural control of cardiovascular reflexes have been reviewed by Korner (1971).

Sensory innervation is not restricted, however, to these regions, but extends into the arterioles. The arteriolar flare that occurs in the triple responses in the skin and conjunctiva depends upon the existence of myelinated sensory fibres that send collaterals that terminate in endings on the vessel walls as well as in the superficial layers of the skin. These collaterals supply the pathway for the axon reflex that produces arteriolar dilatation (Lewis, 1927). The capillaries apparently receive no myelinated fibres, but Landis (1929) noted that a slight pricking sensation was sometimes felt when a microcannula was entering nailfold capillaries in human subjects. Some afferent nerve fibres occur in the walls of veins (Franklin, 1928).

The blood vessels of the skin of the extremities show a distinct vasodilatation after about 5–10 min exposure to cold. This phenomenon is independent of sympathetic innervation but according to Lewis (1929) disappears when the spinal mixed nerves supplying such regions degenerate. He considered that this reaction was mediated by the axon reflex mechanism via the sensory nerves. A similar response to cooling occurs in the rabbit's ear vessels but differs from that in the human skin in being largely independent of innervation (Grant, Bland and Camp, 1932). In both cases the temperature rise and characteristic 'hunting' that occurs in the cooled part are considered to be due to changes in blood flow through the numerous arterio-venous anastomoses in these regions. The cutaneous vasodilatation that occurs on antidromic stimulation of a spinal nerve is considered to be due to the release of H-substance at the nerve plexuses in the skin which acts to produce local vasodilatation (Lewis, 1927). An extremely close relationship between the perivascular nerve plexuses and the spinal nerves has been revealed in the cremaster muscle by Grant (1966). He found that abdominal sympathectomy had no effect on the morphology of these plexuses but that degeneration occurred in them following experimental division of the spinal nerve supplying the muscle.

7

VESSELS AS FUNCTIONAL UNITS

In the preceding chapters the approach has been largely anatomical in that vessels have been considered in terms of their component parts. In this final chapter certain integrated actions of the various types of vessel will be discussed.

THE ARTERIES

THE ELASTIC ARTERIES

These are the largest arterial vessels in the body and comprise the aorta and pulmonary artery and their major branches. These vessels have the highest velocity of flow in the circulation, which, combined with their great luminal diameters, renders them most prone to develop turbulent flow. In addition, in the systemic elastic arteries, their large diameters and wide pulse pressures combine to produce phasically varying intramural tensions exceeding those in any other vessels. It is generally tacitly assumed that the tunica media of these vessels, which have highly characteristic structures formed by alternating elastic and fibromuscular laminae, supply the major mechanical strength of their walls. The tunica adventitia is usually thought of as supplying the site for the ramification and access of the vasa vasorum, the perivascular nerves and the lymphatic plexuses. In addition it forms a fibrous tissue sleeve that tethers the vessel into position whilst allowing it some mobility. These essentially ancillary functions of the adventitia have overshadowed its contribution to the total mechanical properties of these vessels. I would like to make an analogy with the properties of a rubber bladder within a string net. In such a system the rubber bladder, representing the media, exhibits the desirable qualities of rapid compliance and distensibility. The degree of distension obtainable is determined by the dimensions of the enclosing net, or adventitia (figure 62). Once the fibres of the net are put on stretch the properties of the system alter profoundly, from those of the rubber bladder to those of the enclosing string network. I believe that the tunica media of elastic arteries provides a specialized lining to the wall that is able rapidly to adapt through its compliance and plasticity to the continuously altering pressure-flow conditions existing within the lumen of these vessels.

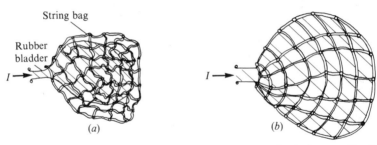

Figure 62. Diagram to illustrate the relation between the intima–media (rubber bladder) and adventitia (string net) of a vessel. (*a*) The normal physiological situation when the blood vessel wall shows rapid compliance and adaptation, the properties of the rubber bladder, to altering pressure–volume conditions. (*b*) The situation of maximal dilatation when the fibres of the adventitia are put on stretch. The response of the system to increasing pressure now changes to those of the string bag, namely low extensibility and high mechanical strength. *I* represents variable input to bladder.

The tunica adventitia, with its massive tendon-like collections of collagen fibrils provides the ultimate physical strength of the wall.

The obliquely oriented muscle cells of the media run diagonally at small angles between adjacent pairs of concentric elastic laminae, giving the impression of the leaflet arrangement within an iris diaphragm. By increasing the angle that the cells make with their adjacent elastic laminae the separation of the laminae can be increased, with the maximum separation occurring when the muscle cells are oriented as nearly as possible radially (figure 63). Since this effect on each pair of the concentric laminae is cumulative very significant changes in the medial thickness can be produced quite readily. There are certain reports in the literature that support this hypothetical mechanism. Wolinsky and Glagov (1964) found in rabbit aortae fixed at various distending pressures that the orientation and shape of the medial muscle cells varied greatly with the distending pressure used. At hydrostatic pressures equal to, or greater than, the diastolic blood pressure the cells were more elongated and were arranged helically between closely opposed pairs of elastic laminae. In aortae fixed at pressures below diastolic, the cells were oriented obliquely to perpendicularly with respect to the adjacent elastic laminae which were much more widely separated. A similar alteration in the orientation of the medial cells spanning the interlaminar spaces has been described by Keech (1960*b*). She found that progressive changes occurred in aortae of lathyritic rats. These included an increasing thickness of the wall with wider interlaminar spaces and the development of a radial orientation of the muscle cells of the media. It is interesting that these changes were found in an experimental condition that results in the production of abnormal collagen with a lowered tensile strength. These alterations of the media may in fact be homeostatic in the sense that they tend to maintain

The arteries 143

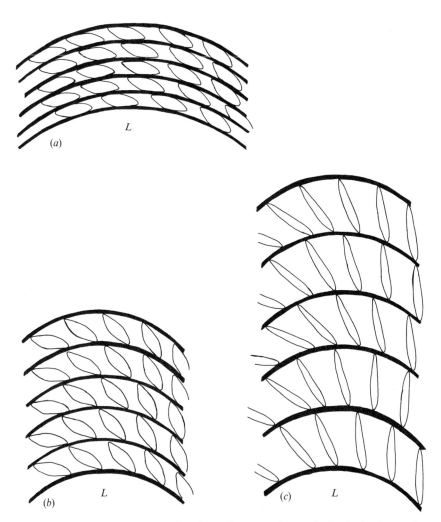

Figure 63. A diagram illustrating how the muscular and elastic laminae of an elastic artery might work in the manner of a series of concentric iris diaphragms to produce rapid local alterations in wall thickness. Such changes could occur passively as the direct result of local alterations in hydrostatic pressure in the lumen (L) and would be distinct from the overall calibre changes produced by muscular contraction.
(a) represents the media under normal physiological conditions.
(b) shows the effect of increasing the angle of the muscle cells to the elastic laminae by about 30°.
(c) shows the combined effect of further increasing the angle of the muscle cells to about 50° and allowing a 30% increase in the length of the muscle fibres.

the lumen diameter constant in the presence of a progressively stretching collagenous framework of the wall.

Other evidence is the recognition by fluoroscopic techniques of large alterations in the diameter of the aortic lumen which are of the order of 20% of the diameter with physiological pulse pressures. Direct observations of exposed aortae show only a 1–4% alteration in external diameter with normal pulse pressure fluctuations. I suggest that, in fact, the media may help to produce the diameter changes of the lumen by its concentric 'iris diaphragm' mechanism within the supporting tunica adventitia. It should not prove too difficult to perform in-vivo microscopy of the walls of elastic arteries in small mammals in order to observe the physiological activity of the smooth muscle and elastic laminae. It may well emerge that the media is performing a much more delicate and subtle role than is usually envisaged within the walls of the elastic arteries in exactly tailoring the internal dimensions and configuration of the arteries to their immediate haemodynamic requirements which can alter from heartbeat to heartbeat and moment to moment in the eventful life of an animal.

THE MUSCULAR ARTERIES

These vessels arise from the elastic arteries and with subsequent branching distribute the arterial blood to the various regions and organs of the body. Hence one speaks of the coronary, the renal, the splenic, the hepatic artery, and so on. They are essentially conduits which are capable of responding physiologically to altered volume flows within their lumina associated with variations in the blood flow within the regions they supply. In this regard the smooth muscle within their walls shows a considerable degree of autoregulatory activity (Lie, Sejersted and Kiil, 1970). Their rich sympathetic innervation mediates their contribution to integrated cardiovascular reflexes, as for instance in the 'fight-or-flight response'. The elastic tissue in their walls is mainly concentrated within the internal elastic lamina, which forms a sharp boundary between the intima and media, and usually to a lesser extent within the external elastic lamina which separates the media and the adventitia. Only small amounts of collagen and elastic tissue are present between the muscle cells of the media. The internal elastic lamina has the same fenestrated sheet-like structure as the laminae within elastic arteries, whilst the external elastic lamina appears more as a condensation of discrete elastic fibrils. In conventionally excised and fixed muscular arteries the internal elastic lamina has a wavy scalloped appearance, with the overlying intima thrown into folds in relation to it. Whilst this appearance is doubtless frequently due to vascular contraction occurring on excising and fixing the vessel, it also probably exists at times during life. Such folding may be detected in physiologically constricted arteries (Van Citters, Wagner and Rushmer,

1962a, b). Burton (1951) has drawn attention to the possible function of elastic tissue in muscular arteries in allowing graded rather than all-or-none vasomotor responses to occur.

ARTERIOLES

These vessels arise as branches of the small muscular arteries, from which they can be distinguished by the absence of an internal elastic lamina. All studies of the microcirculatory system reveal continuous vasomotor activity due to the opening and closing of arterioles so that at no time is the potential cross-sectional area of this section of the circulatory system being fully utilized. Their vasomotor activity is controlled both by the sympathetic nervous system that supplies their walls and also by the local metabolic demands of the tissues they supply. A variety of agents that may act as mediators of this local tissue control have been identified. These vessels provide the major site of resistance within the circulation and therefore are of major importance in the control of the blood pressure. When one considers their relatively simple walls with an endothelial lining invested by from one to three concentric layers of spirally arranged smooth muscle cells (Rhodin, 1967) it might at first sight seem strange that they are able to exercise so much control over their diameters in the presence of relatively high intraluminal pressures. But, because their diameters are so small the tangential stresses within their walls (by the Laplace equation) are very low (Burton, 1951).

METARTERIOLES

These extremely fine vessels are the final arborizations of the arterial system that supply the nutrient-net vessels of the tissues. They may be recognized *in vivo* by their nearly perpendicular origins from the muscular arterioles, by their very direct, almost straight, courses within the tissues, and by the absence of smooth muscle cells within their wall. The velocity of flow within these vessels is higher than within the nutrient-net vessels and because of their extremely narrow calibre the erythrocytes passing through them show parachute or bullet-shaped deformation. Smooth muscle sphincters situated at their origins from the arterioles control their rate of flow and can shut off flow completely within these vessels.

ARTERIO-VENOUS ANASTOMOSES (AVAs)

These structures are readily recognizable on in-vivo observation of the microcirculation where arterioles, or sometimes metarterioles, flow abruptly into a venule or a small vein, which is locally dilated into a fusiform siphon to receive the arterial blood. On entering the venous

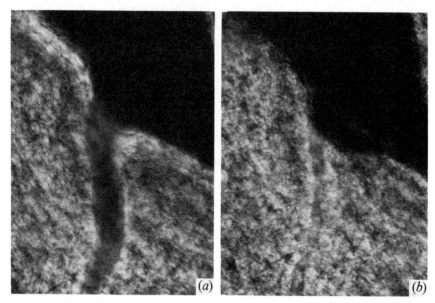

Figure 64. (a) and (b) are photomicrographs, taken within a short interval of one another, showing vasomotion in an arterio-venous anastomosis. An arteriole with a muscular wall enters across the lower margin of the figure and flows directly into the venule at the upper right of the figure. Rabbit ear chamber. Magnification × 180.

siphon the velocity of flow slows almost instantaneously to conform with the general venous velocity of flow. Such AVAs can be seen opening and closing during the period of observation (figure 64).

Histologically the arterial limb of such a structure is characterized by the unusual stellate shape of its lumen. There is plentiful smooth muscle in its wall and elastic tissue is poorly developed. The venous limb is devoid of muscle cells and its wall is merely a fibro-elastic tube enclosing the endothelium. The peculiar structure of the AVA has been linked to the all-or-none character of its activity in the circulation where it is either 'on' or 'off', with little or no graded response possible (Burton, 1954).

AVAs are particularly numerous in the dermis where they control heat dissipation through the skin. When open they increase the volume rate of flow within the dermis to levels greatly in excess of that which could possibly occur through the nutrient vessels that serve the metabolic needs of the skin (Krogh, 1959). A large volume of blood flowing close to the body surface serves to dissipate excess heat to the exterior. When it is necessary to conserve heat the AVAs close down without compromising the metabolism of the skin. Other sites where AVAs are considered to play important roles are the liver, the kidney and possibly also the intestine.

MICROVASCULAR NETWORKS

It is through the myriads of simple minute vessels that form the microcirculatory networks that the cardiovascular system actually performs its essential functions within the tissues. These would include the supply of oxygen and of nutriments, the removal and transport of various products of metabolism, and the distribution of hormones. It is not surprising therefore that such vessels should have been studied in great detail for many years. In addition to histological and electron microscopical techniques, and those of preparing injected specimens for corrosion studies or examination in cleared tissues, a large number of in-vivo techniques have been developed. These utilize such systems as epi-illumination of nailfold, conjunctival and retinal vessels, and quartz rod transillumination of suitably thin regions of certain organs (lung, liver) or naturally occurring membranes such as the hamster's cheek pouch. Direct transillumination can be performed on the frog's foot web and retrolingual membrane, the bat's wing and the mesenteries of various animals. In addition, there have been developed various surgical techniques for implanting windows of transparent materials which permit long-term microscopic observations of the microcirculation using epi-illumination in such sites as the meninges and the inner ear. Descriptions of these various methods may be found in the *First Conference on Microcirculatory Physiology and Pathology* (1954).

Transparent chambers have been developed that permit long-term transillumination microscopy of the mammalian circulation. These derived initially from the histological observation that new blood vessels and fibrous tissue grew into a narrow space separating two glass coverslips implanted within mammalian tissue. Sandison (1924), working in E. R. Clark's laboratory, developed a transparent chamber that could be implanted within the rabbit's ear and be placed upon the stage of a normal compound microscope for study by transmitted light (figure 65). Rabbit ear chambers have evolved rapidly and techniques for the establishment of minute grafts of various organs and tumours in these chambers have been developed. Williams (1954) has given an excellent account of the techniques involved in the use of the rabbit ear chamber. Such chambers can also be installed in surgically created skin flaps of mammals as, for example, in the mouse and rat back and also in the hamster's cheek.

It is extremely important to realize that there is no such thing as a typical stereotyped microcirculatory pattern. Each organ or tissue within the body has a microvascular architecture which is intimately related to its histological structure and which is in its own right just as characteristic. This is immediately obvious if one considers the microvasculature of such organs as the kidney (glomerular tufts, afferent and efferent arterioles,

148 Vessels as functional units

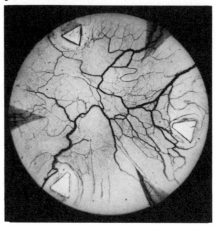

Figure 65. A low-power photomicrograph of a rabbit ear chamber. The networks formed by the nutrient vessels and the venules can be readily identified. The arterioles supplying these vessels are so narrow as to be almost impossible to identify in this figure. Magnification × 9.

vasa recta), the liver (sinusoids) and skeletal muscle (very long nutrient vessels aligned with the muscle cells) (figure 66).

The microcirculation is in a continual state of flux, with any one vessel showing frequent variations in direction and rate of flow and in volume of contained blood. These changes occur in response to the dictates of local tissue requirements but are also generally subject to the overall controls of nerve- and hormone-mediated regulation.

The basic concept of Starling's hypothesis relates the physiological permeability of nutrient vessels to the effect of the difference in the hydrostatic pressures of the blood and the extravascular tissue fluid being opposed by the difference in the colloid osmotic pressures (COP) of the blood and the extravascular tissue fluid (Starling, 1896). When the difference in hydrostatic pressure exceeds the opposing effect of the difference in COP then the net effect will be to provide outward filtration from blood to tissue. Such conditions are envisaged as pertaining to the high pressure (arterial) ends of the nutrient vessels. When the hydrostatic pressure difference is less than the COP difference then the net effect will be to provide inward filtration of fluid and solutes from the tissues into the blood vessels. These conditions are considered to pertain to the venous ends of the nutrient vessels. Experimental observations on individual nutrient vessels and COPs confirm this basic concept (Landis, 1929; Landis and Pappenheimer, 1963). However, it has been pointed out by a number of workers that the concept is too simplistic to fit with the highly variable pressure-flow conditions that are observed within the microcirculation (Landis, 1926, 1929; Nicoll and Webb, 1955; Nicoll and Frayser, 1967; Zweifach, 1971a, b). It is suggested that in fact certain

Microvascular networks 149

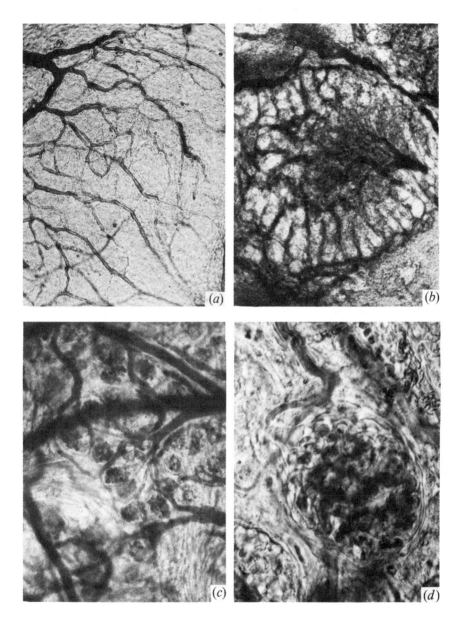

Figure 66. The characteristic vascular patterns of tissues grafted into rabbit ear chambers. (a) smooth muscle, magnification ×70; (b) lung, magnification ×35; (c) kidney, magnification ×150; (d) renal glomerulus, magnification ×400. (c) and (d) reproduced through the kindness of Dr J. B. Hobbs.

vessels with rapid flow and high hydrostatic pressures have net outward filtration over their entire lengths while other vessels, often quite nearby, with slow flow and low hydrostatic pressures, have net inward filtration over their entire lengths. Hence, the high degree of observed vasomotion in the microcirculation is probably geared to the local demands of the tissues so that changes within the micro-environment can dictate to a large degree whether a certain vessel within the tissues is providing or absorbing fluid and solutes at any particular moment.

Due to the extreme simplicity of the walls of nutrient vessels, and the failure to demonstrate effector nerve endings in them, the vasomotor changes observed therein are generally considered to be secondary to changes in the vessels supplying and draining them. Alterations that do occur as the direct effect of changes in their walls are due to reactions of the lining endothelial cells. Agents capable of directly modifying the properties of their endothelium may be listed as being mechanical, thermal, ionizing radiation, chemical, hormonal, immunologic and infectious. The changes that occur are stereotyped. One is increased permeability, which in the presence of adequate blood flow leads to oedema, as in the characteristic weal of the triple response (Lewis, 1927). When the damage is very severe, rapid loss of water and solutes through the vessel wall results in the lumen becoming completely clogged with compact masses of blood cells. This constitutes vascular stasis (Landis, 1927; Krogh, 1959). This state may be reversed on recovery of the endothelial damage, provided tissue necrosis has not supervened. Extremely severe damage can lead to actual disruption of the vessel walls resulting in haemorrhage and thrombosis. The other change is the margination, or 'sticking', of leukocytes on the endothelium and their emigration from the vessels into the surrounding tissues. This phenomenon occurs in the venules of the microvascular networks (Clark and Clark, 1935; Krogh, 1959; Marchesi, 1961; Hurley, 1964; Moses et al., 1964; Cliff, 1966) (figure 67). Exudation of fluid and solutes, including proteins, and the emigration of leukocytes, when taken in conjunction with increased blood flow to the region, form the acute inflammatory reaction.

The changes responsible for the firm adherence of leukocytes to the endothelial membrane are obscure. It seems most probable that an alteration occurs in the vascular endothelium as a result of injury which makes it 'sticky' for leukocytes (Clark and Clark, 1935; Harris, 1960; Grant, Palmer and Sanders, 1962; Cliff, 1966). The observation that leukocytes may form a layer 2–3 cells in thickness on the endothelium indicates that leukocytes are also sticking to leukocytes in inflammation (Allison, Smith and Wood, 1955; Krogh, 1959; Cliff, 1966) and suggests that some plasmatic alteration may be responsible for their sticking. The involvement of the blood clotting system in this phenomenon has been excluded

Microvascular networks 151

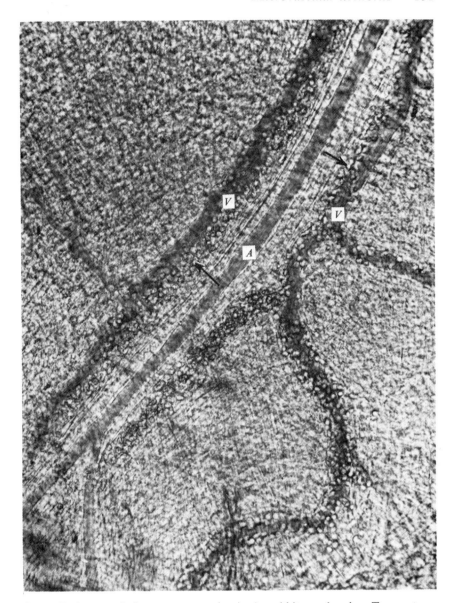

Figure 67. An acute inflammatory reaction in the rabbit ear chamber. Two systems of venules (*V*) flank an arteriole (*A*). There is severe sticking of leukocytes in the venules, giving to their walls a cobblestone paving appearance. Note the highly asymmetric pattern of leukocyte-sticking in regions (arrows). Magnification × 250.

(Allison and Lancaster, 1960; Grant et al., 1962). The changes responsible for leukocyte sticking can occur within a matter of 30 seconds of direct damage to the vessel wall (Clark and Clark, 1935) but with less severe stimulation a latent period of $1-1\frac{1}{2}$ hours may be noted before leukocyte sticking occurs (Grant et al., 1962; Moses et al., 1964; Graham et al., 1965; Cliff, 1966). This frequently noted time lag could be interpreted as the time required for the adequate build-up of chemical mediators of inflammation formed within the tissues as a result of injury (Allison et al., 1955; Buckley, 1963). A specific chemical entity that is responsible for mobilizing leukocytes in the tissue has been proposed by Menkin (1955, 1960) and called 'leukotaxine'. However, 'leukotaxine' is not a single compound but is most probably composed of a variety of protein derivatives (Moon and Tershakovec, 1953). There is evidence that some plasma component(s) may in fact be activated to produce leukocyte emigration by contact with injured tissue (Hurley, 1963). Graham and others (1965) investigated the role of activated Hageman factor in this context and excluded its activity as a kinin releaser by comparing its effects with those of bradykinin. Harris (1954, 1960), in reviewing the subject of chemotactic influences on leukocytes, concluded that their emigration from blood vessels probably was not due to any direct chemical effect upon the leukocytes themselves. The mobilization of leukocytes does not occur as readily in some sites as others. Hence Rocha and Fekety (1964), using thermal injury, established that the renal medulla showed a deficient mobilization of granulocytes when compared with the renal cortex, and suggested that this deficiency played a part in the higher susceptibility of the renal medulla to bacterial infection. Landis (1964) demonstrated that T1824 (Evans blue) dye showed a prominent spotty passage through venule walls. Whilst the interpretation of his results regarding the physiological passage of large molecules is not tenable in light of subsequent work on T1824–albumin binding studies (Levick and Michel, 1973), his suggestion that these physiological pathways may provide the routes of migration for leukocytes is worthy of consideration.

Whilst the accumulation of leukocytes within the tissues is very important from the pathological point of view there are other classical signs of inflammation that are of great clinical value. These are local redness, increased heat, and swelling, all three signs being due to alterations in the blood vessels. The redness and heat are signs of increased blood flow in the inflamed region; this increase is due to the opening up of local arterioles. More generalized arteriolar changes in the skin are responsible for the flare of the triple response (Lewis, 1927). There are certain features of the triple response that should be stressed. Firstly, it is a phenomenon occurring in the skin and conjunctiva, secondly, it requires the mediation of intact (probably sensory) nerve axons to develop fully, and thirdly, the chemical agent released from the damaged cells of the epidermis or dermis is very properly designated 'H-substance' (Lewis, 1927, 1929). Whilst Lewis found many similarities between 'H-substance' and histamine he pointed out that similar cutaneous responses could also be obtained by pricking-in atropine or morphine. In addition to the responses mediated through stimulation of the sensory axon reflexes, the direct effects of histamine upon blood vessels have been widely studied (Spector and Willoughby, 1964). Responses of peripheral vessels to histamine are difficult to interpret and vary from site to site and species to species (Dale, 1920*a*; Krogh, 1959; Greenway and Stark, 1971). Whilst histamine generally produces capillary dilatation (Dale, 1920*b*; Lewis, 1927; Altura and Zweifach, 1965; Altura, 1970; Northover and Northover, 1970) it should be noted that Buckley and Ryan (1969) found no vasomotor changes when histamine was applied topically to capillary vessels within the rat's mesentery.

Bradykinin would appear to be a more promising candidate to fill the local hormone role, initially proposed for histamine, of an endogenous vasodilator (Vane, 1969). In addition, it has been suggested that bradykinin may act as a chemical mediator of the inflammatory response (Elliott, Horton and Lewis, 1960; Lewis, 1964). This nonapeptide is a potent vasodilator, increases capillary permeability, and in higher concentrations can produce leukocyte migration. In human subjects this material causes a sensation of pain, which is a feature of inflammation, whereas histamine produces a pricking or itching sensation.

Serotonin (5-hydroxytryptamine, 5HT) has also been studied as a possible chemical mediator of the inflammatory reaction (Spector and Willoughby, 1964). In some species, particularly rodents, this material increases vascular permeability. In human subjects, whilst being more potent than bradykinin in producing pain, it has an epinephrine-like action in causing vasoconstriction in the skin and vasodilatation in muscles (Lewis, 1964).

The mechanism responsible for the increased vascular permeability that occurs in inflamed tissues has attracted a good deal of attention. This altered permeability allows large amounts of protein and fluid to leave the blood vessels and collect in the tissues, resulting in local swelling and oedema (Yoffey and Courtice, 1970). Such increased permeability can be detected macroscopically in various skin test procedures by the local accumulation, following intravenous injection, of a colloidal dye such as trypan blue or of a dye such as T1824 (Evans blue) that complexes with albumin. For local oedema to develop two criteria must be met: firstly there must be an alteration in the endothelial membrane allowing increased passage of colloids and secondly there must be an adequate rate of blood flow through the region (Lewis, 1927; Northover and Northover, 1970). It is not correct to equate the changes of increased vascular permeability with the inflammatory reaction as a whole, as unfortunately tends to occur.

The migration of leukocytes, which is a highly significant feature of this pathological process, can be dissociated quite clearly from the phenomenon of increased vascular permeability in terms of the types of vessel involved (Cotran and Majno, 1964) and the mediators which provoke the particular phenomenon (Hurley, 1964). Clark and Clark (1935) were able to observe vessels *in vivo* that showed evidence of increased permeability with no concomitant leukocyte sticking occurring on their endothelium, and Moon and Tershakovec (1953) found no correlation between the degree of leukocyte migration and the degree of permeability changes that were detected in blood vessels.

The increased permeability of blood vessels following thermal injury shows a biphasic reaction. There is an immediate response within venules which is chemically mediated. This is followed after an interval varying from 15 minutes to 4 hours by a delayed response which involves small venules and capillaries and is due to the direct damage inflicted upon the walls of these vessels by the stimulus (Cotran and Majno, 1964; Cotran, 1967; Hurley, Ham and Ryan, 1967).

A very significant and constant feature found on electron microscopic examination of vessels which have become abnormally permeable to fluid and colloids is the presence in the endothelial membrane of gaps produced by localized widening and separation of the endothelial cell junctions (Majno, 1970). This change is found in venules within areas of acute inflammation (Marchesi, 1962; Cotran, 1967) and allergic inflammation (Movat and Fernando, 1963) and is also found within the vessels of the ciliary body in the presence of a corneal fistula (Pappas and Tennyson, 1962) and within the cerebral arterial vessels of hypertensive animals (Giacomelli et al., 1970). Similar intercellular gaps are found in the endothelium of venules after treatment with such possible chemical mediators of the inflammatory reaction as histamine, serotonin and bradykinin in the endothelium of venules (Majno and Palade, 1961; Rowley, 1964; Majno et al., 1969). Constantinides and Robinson (1969a) found that these compounds had the same effect upon the endothelium of arteries. The gaps in venular endothelium occur most commonly where three endothelial cells form a junction (Majno and Palade, 1961), a view initially held, but subsequently discounted, by Krogh (1959). It has been proposed that gaps are produced at the endothelial cell junctions by a smooth muscle-like contraction of the cell cytoplasm of abnormally permeable endothelium (Constantinides and Robinson, 1969a, b; Majno et al., 1969; Giacomelli et al., 1970; Northover and Northover, 1970). Quite well-developed tracts of fine fibrils strongly reminiscent of the myofibrils of smooth muscle cells have also been identified within endothelial cells from arteries, arterioles, capillaries and venules by other workers (Hama, 1961; French, 1963; Terzakis, 1963; Ancla and de Brux, 1964; Phelps and Luft, 1969; Gerrity and Cliff, 1972) (figure 30). Fluorescein-conjugated antibodies to actomyosin have been used successfully to stain endothelial cells (Becker and Murphy, 1968; Becker, 1970). Majno (1970) has inferred from this that the fibrillar tracts detected electron microscopically within endothelial cells are in fact actomyosin. I feel that this is not yet fully established since other workers have failed to demonstrate the existence of actomyosin within endothelial cells using antibody techniques (Knieriem *et al.*, 1968). Other aspects of endothelial contractility have been discussed in chapter 3.

The fact that these permeability changes are localized to the venules within the microcirculation, even though the endothelial cells of other minute vessels show evidence of contraction, seems to indicate that the endothelial junctions of venules are either initially more readily pulled apart or else are preferentially modified by the mediators of the inflammatory reaction (Majno *et al.*, 1969).

POST-CAPILLARY VENULES OF LYMPHOID TISSUE

The post-capillary venules located within the extrafollicular regions of the cortex of lymph nodes and within the Peyer's patches of the intestine are major sites for the physiological migration of small lymphocytes from blood to lymph (Gowans and Knight, 1964). In any other vessels the identification of leukocytic migration through endothelium is an incontrovertible sign of inflammation. These specialized venules can be identified by their large calibre, by their cuffing with numerous small lymphocytes within their paravascular sheaths and generally by their 'high' endothelium (Sainte-Marie, 1966; Schoefl, 1970, 1972; Claesson *et al.*, 1971; Mikata and Niki, 1971) (figure 45). Their high endothelium is composed of readily deformable cuboidal or even columnar endothelial cells. These cells are apparently able to selectively produce margination of the small lymphocytes and to allow their migration from the vessels by way of their intercellular junctions (Schoefl, 1970, 1972; Claesson *et al.*, 1971). The intercellular junctions of this endothelium are permeable to intravenously injected particles of carbon, ferritin (Schoefl, 1970) and thorium dioxide (Mikata and Niki, 1971). The interesting question as to whether the high endothelial cells are actually formed in response to the large traffic of lymphocytes through the vessel walls has been discussed by Schoefl (1972). The role these vessels have in selectively secreting lymphocytes suggests to me the term 'adenoidal venules' rather than the long and cumbersome term 'high endothelial post-capillary venules'.

THE VEINS

In contrast to arterial vessels, veins have a low ratio of wall thickness to lumen diameter. They supply the large capacitance–low pressure limb of the circulation. It is wrong, however, to consider all such vessels as being merely passive conduits returning blood to the heart. Rhythmic contractions of veins in the bat's wing as well as of the mesenteric and portal veins in a number of mammals aid the venous flow in such regions. This action is probably analogous to that of the peripheral hearts of primitive vertebrates.

In response to altering pressure and volume flow conditions occurring within their lumina the smooth muscle in their walls can produce venoconstriction or venodilatation. However, once again in contrast to arteries, veins are also able to modify their cross-sectional area by altering the shape of their lumina. When volume flow is minimal the thin-walled deformable vein is flattened and the lumen becomes a ribbon-like slit. As the volume flow increases so the lumen progressively becomes more elliptical and finally circular.

The poorer development of elastic tissues within their walls can be related to the lower levels of wall stress in veins as compared to arteries. The surprising strength of vein walls is due to their well-developed collagenous components. This strength is graphically illustrated by the successful grafting of venous implants into arteries without subsequent development of aneurysms.

In general, veins are more variable in their anatomical distribution than are the corresponding arteries.

REFERENCES

Aars, H. (1971). Effects of altered smooth muscle tone on aortic diameter and aortic baroreceptor activity in anaesthetized rabbits. *Circulation Res.*, **28**, 254–262.

Aars, H. and Solberg, L. A. (1971). Effect of turbulence on the development of aortic atherosclerosis. *Atherosclerosis*, **13**, 283–287.

Aarseth, P. (1970). Reduction in pulmonary blood volume after a blood loss. *Acta physiol. scand.*, **80**, 459–469.

Abraham, A. (1969). *Microscopic Innervation of the Heart and Blood Vessels in Vertebrates including Man.* Pergamon Press, Oxford.

Adams, C. W. M., Bayliss, Olga B. and Ibrahim, M. Z. M. (1963). The distribution of lipids and enzymes in the aortic wall in dietary rabbit atheroma and human atherosclerosis. *J. Path. Bact.*, **86**, 421–430.

Adams, C. W. M., Bayliss, Olga B., Ibrahim, M. Z. M. and Webster, M. W., Jr (1963). Phospholipids in atherosclerosis: the modification of the cholesterol granuloma by phospholipid. *J. Path. Bact.*, **86**, 431–436.

Adams, C. W. M., Bayliss, Olga B., Davison, A. N. and Ibrahim, M. Z. M. (1964). Autoradiographic evidence for the outward transport of ^{3}H-cholesterol through rat and rabbit aortic wall. *J. Path. Bact.*, **87**, 297–304.

Adelson, E., Heitzman, E. J. and Fennessey, J. F. (1954). Thrombo-hemolytic thrombocytopenic purpura. *A.M.A. Archs Int. Med.*, **94**, 42–60.

Aikawa, M. and Koletsky, S. (1970). Arteriosclerosis of the mesenteric arteries of rats with renal hypertension. *Am. J. Path.*, **61**, 293–304.

Akester, A. R. (1967). Renal portal shunts in the kidney of the domestic fowl. *J. Anat.*, **101**, 569–594.

Akester, A. R. and Mann, S. P. (1969). Adrenergic and cholinergic innervation of the renal portal valve in the domestic fowl. *J. Anat.*, **104**, 241–252.

Akmayev, I. G. (1971). Morphological aspects of the hypothalamic–hypophyseal system. II. Functional morphology of pituitary microcirculation. *Z. Zellforsch. mikrosk. Anat.*, **116**, 178–194.

Albert, E. N. and Bhussry, B. R. (1967). The effects of multiple pregnancies and age on the elastic tissue of uterine arteries in the guinea pig. *Am. J. Anat.*, **121**, 259–269.

Alexander, A. F. and Jensen, R. (1963). Normal structure of bovine pulmonary vasculature. *Am. J. vet. Res.*, **24**, 1083–1093.

Allison, A. C., Davies, P. and De Petris, S. (1971). Role of contractile microfilaments in macrophage movement and endocytosis. *Nature (New Biol.)*, **232**, 153–155.

Allison, F., Jr and Lancaster, M. G. (1960). Studies on the pathogenesis of acute inflammation. II. The relationship of fibrinogen and fibrin to the leucocytic sticking reaction in ear chambers of rabbits injured by heat. *J. exp. Med.*, **111**, 45–64.

Allison, F., Jr, Smith, Mary R. and Wood, W. B. (1955). Studies on the pathogenesis of acute inflammation; the inflammatory reaction to thermal injury as observed in the rabbit ear chamber. *J. exp. Med.*, **102**, 655–668.

Altura, B. M. (1970). Contractile responses of microvascular smooth muscle to antihistamines. *Am. J. Physiol.*, **218**, 1082–1091.
Altura, B. M. and Zweifach, B. W. (1965). Antihistamines and vascular reactivity. *Am. J. Physiol.*, **209**, 545–549.
Ancla, M. and Brux, J. de (1964). Étude au microscope électronique des artérioles spiralées de l'endomètre humain. *Ann. Anat. Path.*, **9**, 209–222.
Andrew, W. (1962). Life span of cell and organism. In *Biological Aspects of Aging*, ed. N. W. Shock. Columbia University Press, N.Y. and London. pp. 123–130.
Anitschkow, N. N. (1967). A history of experimentation on arterial atherosclerosis in animals. In *Cowdry's Arteriosclerosis, a Survey of the Problem*, 2nd editn. ed. H. T. Blumenthal. C. Thomas, Springfield, Illinois. pp. 21–44.
Arey, L. B. (1936). Wound healing. *Physiol. Rev.*, **16**, 327–406.
Arey, L. B. (1963). The development of peripheral blood vessels. In *The Peripheral Blood Vessels*, ed. J. L. Orbison and D. E. Smith. Williams and Wilkins, Baltimore. pp. 1–16.
Arndt, J. O., Stegall, H. F. and Wicke, H. J. (1971). Mechanics of the aorta *in vivo*. *Circulation Res.*, **28**, 693–704.
Aschoff, L. (1933). Introduction to *Arteriosclerosis – A Survey of the Problem*, ed. E. V. Cowdry. Publication of Josiah Macy Jnr Foundation, Macmillan Co., N.Y.
Ashford, T. P., Freiman, D. G. and Weinstein, M. C. (1968). The role of the intrinsic fibrinolytic system in the prevention of stasis thrombosis in small veins. *Am. J. Path.*, **52**, 1117–1127.
Astrup, T. and Buluk, K. (1963). Thromboplastic and fibrinolytic activities in vessels of animals. *Circulation Res.*, **13**, 253–260.
Bahr, G. F. (1954). Osmium tetroxide and Ruthenium tetroxide and their reactions with biologically important substances. *Expl Cell Res.*, **7**, 457–479.
Bainton, D. F. and Farquhar, M. G. (1970). Segregation and packaging of granule enzymes in eosinophilic leukocytes. *J. Cell Biol.*, **45**, 54–73.
Baldwin, D., Taylor, C. B. and Hass, G. M. (1950). A comparison of arteriosclerotic lesions produced in young and in old rabbits by freezing the aorta. *Archs Path.*, **50**, 122–131.
Balis, J. U., Chan, A. S. and Conen, P. E. (1967). Morphogenesis of human aortic coarctation. *Exp. and Mol. Path.*, **6**, 25–38.
Baló, J. and Banga, I. (1950). The elastolytic activity of pancreatic extracts. *Biochem. J.*, **46**, 384–387.
Baradi, A. F. and Hope, J. (1964). Observations on ultrastructure of rabbit mesothelium. *Expl Cell Res.*, **34**, 33–44.
Barber, V. C. and Graziadei, P. (1967). The fine structure of cephalopod blood vessels. III. Vessel innervation. *Z. Zellforsch. mikrosk. Anat.*, **77**, 162–174.
Barer, R., Joseph, S. and Meek, G. A. (1960). The origin and fate of nuclear membranes in meiosis. *Proc. Roy. Soc., B*, **152**, 353–366.
Barnard, W. G. and Robb-Smith, A. H. T. (1945). *Kettle's Pathology of Tumours*. H. K. Lewis and Co., London.
Barr, M. L., Bertram, L. F. and Lindsay, H. A. (1950). The morphology of the nerve cell nucleus, according to sex. *Anat. Rec.*, **107**, 283–297.
Batson, O. V. (1942). The role of the vertebral veins in metastatic processes. *Ann. Int. Med.*, **16**, 38–45.
Bazett, H. C., Love, L., Newton, M., Eisenberg, L., Day, R. and Forster, R. (1948). Temperature changes in blood flowing in arteries and veins in man. *J. Appl. Physiol.*, **1**, 3–19.
Beard, J. W. and Beard, L. A. (1927). The phagocytic activity of endothelium in the embryo chick. *Am. J. Anat.*, **40**, 295–313.
Becker, C. G. (1970). Investigations of cardiovascular contractile proteins. In

Vascular Factors and Thrombosis, ed. K. M. Brinkhous *et al.* Schattauer Verlag, Stuttgart and N.Y. pp. 31–49.

Becker, C. G. and Murphy, G. E. (1968). Demonstration of actomyosin in cells of heart valve endothelium, intima, the arteriosclerotic plaque, and endocardial and myocardial Aschoff bodies. *Am. J. Path.*, **52**, 22 (Abstract).

Becker, C. G. and Murphy, G. E. (1969). Demonstration of contractile protein in endothelium and cells of the heart valves, endocardium, intima, arteriosclerotic plaques, and Aschoff bodies of rheumatic heart disease. *Am. J. Path.*, **55**, 1–37.

Behnke, O. and Zelander, T. (1970). Preservation of intercellular substances by the cationic dye Alcian blue in preparative procedures for electron microscopy. *J. Ultrastruct. Res.*, **31**, 424–438.

Bell, C. (1972). Autonomic nervous control of reproduction circulatory and other factors. *Pharmac. Rev.*, **24**, 657–736.

Bennett, H. S. (1963). Morphological aspects of extracellular polysaccharides. *J. Histochem. Cytochem.*, **11**, 14–23.

Bennett, H. S. and Luft, J. H. (1959). s-collidine as a basis for buffering fixatives. *J. biophys. biochem. Cytol.*, **6**, 113–114.

Bennett, H. S., Luft, J. H. and Hampton, J. C. (1959). Morphological classifications of vertebrate blood capillaries. *Am. J. Physiol.*, **196**, 381–390.

Bennett, T. (1971). The adrenergic innervation of the pulmonary vasculature, the lung and thoracic aorta, and on the presence of aortic bodies in the domestic fowl (*Gallus gallus domesticus* L.). *Z. Zellforsch. mikrosk. Anat.*, **114**, 117–134.

Bennett, T. and Malmfors, T. (1970). The adrenergic nervous system of the domestic fowl (*Gallus domesticus* L.). *Z. Zellforsch. mikrosk. Anat.*, **106**, 22–50.

Bentley, J. P. and Hanson, A. N. (1969). The hydroxyproline of elastin. *Biochim. biophys. Acta*, **175**, 339–344.

Bergel, D. H. (1961). The static elastic properties of the arterial wall. *J. Physiol.*, **156**, 445–457.

Berlepsch, K. von (1970). Polysaccharides of the vessel wall distribution and function. In *Vascular Factors and Thrombosis*, ed. K. M. Brinkhous *et al.* Schattauer Verlag, Stuttgart and N.Y. pp. 51–57.

Berliner, R. W., Levinsky, N. G., Davidson, D. G. and Eden, M. (1958). Dilution and concentration of the urine and the action of antidiuretic hormone. *Am. J. Med.*, **24**, 730–744.

Berry, C. L. (1969). Changes in the wall of the pulmonary artery after banding. *J. Path.*, **99**, 29–32.

Bhisey, A. N. and Freed, J. J. (1971). Amoeboid movement induced in cultured macrophages by colchicine or vinblastine. *Expl Cell Res.*, **64**, 419–429.

Billroth, T. (1856). *Untersuchungen über die Entwicklung der Blutgefässe*. Berlin.

Biological Handbooks (1971). *Respiration and Circulation*, ed. P. L. Altam and D. S. Dittmer. Federation of American Societies for Experimental Biology, Bethesda, Maryland.

Björkerud, S. (1969). Atherosclerosis initiated by mechanical trauma in normolipidemic rabbits. *J. Atheroscler. Res.*, **9**, 209–213.

Blanchette-Mackie, E. J. and Scow, R. O. (1971). Sites of lipoprotein lipase activity in adipose tissue perfused with chylomicrons. Electron microscope cytochemical study. *J. Cell Biol.*, **51**, 1–25.

Bloch, E. H. (1955). The *in vivo* microscopic vascular anatomy and physiology of the liver as determined with the quartz rod method of transillumination. *Angiology*, **6**, 340–349.

Bloom, S. (1970). Structural changes in nuclear envelopes during elongation of heart muscle cells. *J. Cell Biol.*, **44**, 218–223.

Bloom, W. (1930). Studies on fibers in tissue culture: development of elastic fibers in cultures of embryonic heart and aorta. *Arch. exp. Zellforsch*, **9**, 6–13.

Böck, P., Stockinger, L. and Vyslonzil, E. (1970). Die Feinstruktur des Glomus caroticum beim Menschen. *Z. Zellforsch. mikrosk. Anat.*, **105**, 543–568.

Bodenheimer, T. S. and Brightman, M. W. (1968). A blood brain barrier to peroxidase in capillaries surrounded by perivascular space. *Am. J. Anat.*, **122**, 249–268.

Bodian, D. (1970). An electron microscopic characterization of classes of synaptic vesicles by means of controlled aldehyde fixation. *J. Cell Biol.*, **44**, 115–124.

Bohr, D. F., Filo, R. S. and Guthe, K. F. (1962). Contractile protein in vascular smooth muscle. *Physiol. Rev.*, **42**, Suppl. 5, 98–107.

Bolton, T. B. and Nishihara, H. (1970). The fine structure of avian vascular muscle and its electrical constants obtained with intracellular micro-electrodes. *J. Physiol.*, **208**, 20–21P.

Bond, T. P., Derrick, J. R. and Guest, M. M. (1964). Microcirculation during hypothermia; high speed cinematograph studies. *Archs Surg.*, **89**, 887–890.

Bondjers, G. and Björnheden, T. (1970). Experimental atherosclerosis induced by mechanical trauma in rats. *Atherosclerosis*, **12**, 301–306.

Bourdeau, J. E., Carone, F. A. and Ganote, C. E. (1972). Serum albumin uptake in isolated perfused renal tubules. Quantitative and electron microscope radioautographic studies in three anatomical segments of the rabbit nephron. *J. Cell Biol.*, **54**, 382–398.

Boyd, W. (1953). *A Text-Book of Pathology*, 6th editn. Henry Kimpton, London.

Braasch, D. (1971). Red cell deformability and capillary blood flow. *Physiol. Rev.*, **51**, 679–701.

Brash, J. C. and Jamieson, E. B. (eds) (1947). *Cunningham's Manual of Practical Anatomy*, 10th editn. Oxford University Press, London, N.Y., Toronto.

Brenner, R. M. (1966). Fine structure of adrenocortical cells in adult male Rhesus monkeys. *Am. J. Anat.*, **119**, 429–454.

Brookes, M. (1970). Arteriolar blockade: a method of measuring blood flow rates in the skeleton. *J. Anat.*, **106**, 557–563.

Bruns, R. R. and Palade, G. E. (1968a). Studies on blood capillaries. I. General organization of blood capillaries in muscle. *J. Cell Biol.*, **37**, 244–276.

Bruns, R. R. and Palade, G. E. (1968b). Studies on blood capillaries. II. Transport of ferritin molecules across the wall of muscle capillaries. *J. Cell Biol.*, **37**, 277–299.

Buck, R. C. (1958). Fine structure of endothelium of large arteries. *J. biophys. biochem. Cytol.*, **4**, 187–190.

Buck, R. C. (1959). Electron microscopic observations on capillaries of atherosclerotic aorta. *Archs Path.*, **67**, 656–659.

Buck, R. C. (1961). Intimal thickening after ligature of arteries. An electron microscopic study. *Circulation Res.*, **9**, 418–426.

Buck, R. C. (1962). Lesions in the rabbit aorta produced by feeding a high cholesterol diet followed by a normal diet. An electron microscopic study. *Brit. J. exp. Path.*, **43**, 236–240.

Buckley, I. K. (1963). Delayed secondary damage and leucocyte chemotaxis following focal aseptic heat injury *in vivo*. *Exp. and Mol. Path.*, **2**, 402–417.

Buckley, I. K. and Ryan, G. B. (1969). Increased vascular permeability. *Am. J. Path.*, **55**, 329–347.

Bunting, C. H. (1939). New formation of elastic tissue in adhesions between serous membranes and in myocardial scars. *Archs Path.*, **28**, 306–312.

Burch, G. E., Tsui, C. Y. and Harb, J. M. (1971). Pathologic changes of aorta and coronary arteries of mice infected with Coxsackie B_4 virus. *Proc. Soc. exp. Biol. Med.*, **137**, 657–661.

Burkel, W. E. (1970). The fine structure of the terminal branches of the hepatic arterial system of the rat. *Anat. Rec.*, **167**, 329–350.
Burn, J. H. (1971). Release of noradrenaline from sympathetic endings. *Nature*, **231**, 237–240.
Burnstock, G. (1972). Purinergic nerves. *Pharmac. Rev.*, **24**, 509–581.
Burnstock, G. and Merrillees, N. C. R. (1964). Structural and experimental studies on autonomic nerve endings in smooth muscle. In *Pharmacology of Smooth Muscle*, Proceedings of the 2nd International Pharmacological Meeting, Vol. 6. Pergamon Press, N.Y. pp. 1–17.
Burnstock, G., Gannon, B. and Iwayama, T. (1970). Sympathetic innervation of vascular smooth muscle in normal and hypertensive animals. *Circulation Res.*, **27**, Suppl. II, 5–21.
Burri, P. H. and Weibel, E. R. (1968). Beeinflussung einer spezifischen cytoplasmatischen Organelle von Endothelzellen durch Adrenalin. *Z. Zellforsch. mikrosk. Anat.*, **88**, 426–440.
Burton, A. C. (1951). On the physical equilibrium of small blood vessels. *Am. J. Physiol.*, **164**, 319–329.
Burton, A. C. (1954). Relation of structure to function of the tissues of the wall of blood vessels. *Physiol. Rev.*, **34**, 619–642.
Burton, A. C. (1967). Physiologic considerations: hemodynamics as related to structure. In *Cowdry's Arteriosclerosis, a Survey of the Problem*, 2nd editn, ed. H. T. Blumenthal. C. Thomas, Springfield, Illinois. pp. 66–80.
Caen, J. P., Legrand, Y. and Sultan, Y. (1970). Platelet-collagen interactions. In *Vascular Factors and Thrombosis*, ed. K. M. Brinkhous *et al.* Schattauer Verlag, Stuttgart and N.Y. pp. 181–197.
Caesar, R., Edwards, G. A. and Ruska, H. (1957). Architecture and nerve supply of mammalian smooth muscle tissue. *J. biophys. biochem. Cytol.*, **3**, 867–877.
Campbell, W. G., Jr and Santos-Buch, C. A. (1959). Widely distributed necrotizing arteritis induced in rabbits by experimental renal alterations. I. Comparison with the vascular lesions induced by injections of foreign serum. *Am. J. Path.*, **35**, 439–465.
Cappell, D. F. (1929a). Intravitam and supravital staining. I. The principles and general results. *J. Path. Bact.*, **32**, 593–628.
Cappell, D. F. (1929b). Intravitam and supravital staining. II. Blood and organs. *J. Path. Bact.*, **32**, 629–674.
Cappell, D. F. (1958). *Muir's Text-Book of Pathology*, 7th editn. Edward Arnold, London.
Caro, C. G., Fitz-Gerald, J. M. and Schroter, R. C. (1970). Wall shear rate in arteries and distribution of early atheroma. In *Vascular Factors and Thrombosis*, ed. K. M. Brinkhous *et al.* Schattauer Verlag, Stuttgart and N.Y. pp. 111–116.
Caro, L. G. (1961). Electron microscopic radioautography of thin sections: the Golgi zone as a site of protein concentration in pancreatic acinar cells. *J. biophys. biochem. Cytol.*, **10**, 37–44.
Caro, L. G. and Palade, G. E. (1964). Protein synthesis, storage, and discharge in the pancreatic exocrine cell. An autoradiographic study. *J. Cell Biol.*, **20**, 473–495.
Carter, D., Einheber, A., Bauer, H., Rosen, H. and Burns, W. F. (1966). The role of microbial flora in uremia. II. Uremic colitis, cardiovascular lesions, and biochemical observations. *J. exp. Med.*, **123**, 251–265.
Casley-Smith, J. R. (1968). The dimensions and numbers of small vesicles in blood and lymphatic endothelium and in mesothelium. *J. Anat.*, **103**, 202–203 (Abstract).
Caulfield, J. B. (1957). Effects of varying the vehicle for OsO_4 in tissue fixation. *J. biophys. biochem. Cytol.*, **3**, 827–829.

Cauna, N. (1970). The fine structure of the arterio-venous anastomosis and its nerve supply in the human nasal respiratory mucosa. *Anat. Rec.*, **168**, 9–22.

Cavallo, T., Sade, R., Folkman, J. and Cotran, R. S. (1972). Tumour angiogenesis. Rapid induction of endothelial mitoses demonstrated by autoradiography. *J. Cell Biol.*, **54**, 408–420.

Chalkey, H. W., Algire, G. H. and Morris, H. P. (1946). Effect of the level of dietary protein on vascular repair in wounds. *J. natn. Cancer Inst.*, **6**, 363–372.

Chambers, R. and Zweifach, B. W. (1944). Topography and function of the mesenteric capillary circulation. *Am. J. Anat.*, **75**, 173–205.

Chambers, R. and Zweifach, B. W. (1947). Intercellular cement and capillary permeability. *Physiol. Rev.*, **27**, 436–463.

Chan, A. S., Balis, J. U. and Conen, P. E. (1965). Maturation of smooth muscle cells in developing human aorta. *Anat. Rec.*, **151**, 334 (Abstract).

Chiba, C., Wolf, P. L., Gudbjarnason, S., Chrysohou, A., Ramos, H., Pearson, B. and Bing, R. J. (1962). Studies on the transplanted heart – its metabolism and histology. *J. exp. Med.*, **115**, 853–865.

Claesson, M. H., Jørgensen, O. and Røpke, C. (1971). Light and electron microscopic studies of the paracortical post-capillary high-endothelial venules. *Z. Zellforsch. mikrosk. Anat.*, **119**, 195–207.

Clark, E. R. (1936). Growth and development of function in blood vessels and lymphatics. *Ann. Int. Med.*, **9**, 1043–1049.

Clark, E. R. (1946). Intercellular substance in relation to tissue growth. *Ann. N.Y. Acad. Sci.*, **46**, 733–742.

Clark, E. R. and Clark, E. L. (1932). Observations on living preformed blood vessels as seen in a transparent chamber inserted into the rabbit's ear. *Am. J. Anat.*, **49**, 441–477.

Clark, E. R. and Clark, E. L. (1935). Observations on changes in blood vascular endothelium in the living animal. *Am. J. Anat.*, **57**, 385–438.

Clark, E. R. and Clark, E. L. (1939). Microscopic observations on the growth of blood capillaries in the living mammal. *Am. J. Anat.*, **64**, 251–301.

Clark, E. R. and Clark, E. L. (1940). Microscopic observations on the extra-endothelial cells of living mammalian blood vessels. *Am. J. Anat.*, **66**, 1–49.

Clark, E. R. and Clark, E. L. (1943). Caliber changes in minute blood vessels observed in the living mammal. *Am. J. Anat.*, **73**, 215–250.

Clark, E. R. and Clark, E. L. (1947). Microscopic studies on the regeneration of medullated nerves in the living mammal. *Am. J. Anat.*, **81**, 233–262.

Clark, E. R., Clark, E. L. and Williams, R. G. (1934). Microscopic observations in the living rabbit of the new growth of nerves and the establishment of nerve-controlled contractions of newly formed arterioles. *Am. J. Anat.*, **55**, 47–77.

Clark, E. R., Kirby-Smith, H. T., Rex, R. O. and Williams, R. G. (1930). Recent modifications in the method of studying living cells and tissues in transparent chambers inserted in the rabbit's ear. *Anat. Rec.*, **47**, 187–211.

Clark, W. G. and Jacobs, E. (1950). Experimental nonthrombocytopenic vascular purpura: A review of the Japanese literature, with preliminary confirmatory report. *Blood*, **5**, 320–328.

Clarke, J. A. (1965). An X-ray microscopic study of the postnatal development of the vasa vasorum in the human aorta. *J. Anat.*, **99**, 877–889.

Claude, A. (1955). Fine structure of cytoplasm. In *Fine structure of Cells* (Symposium). Noordhoof Ltd, Groningen, Holland. pp. 307–314.

Cliff, W. J. (1963). Observations on healing tissue: a combined light and electron microscopic investigation. *Phil. Trans. B*, **246**, 305–325.

Cliff, W. J. (1965). Kinetics of wound healing in rabbit ear chambers, a time lapse cinemicroscopic study. *Q. J. exp. Physiol.*, **50**, 79–89.

Cliff, W. J. (1966). The acute inflammatory reaction in the rabbit ear chamber with particular reference to the phenomenon of leukocytic migration. *J. exp. Med.*, **124**, 543–556.

Cliff, W. J. (1967). The aortic tunica media in growing rats studied with the electron microscope. *Laboratory Invest.*, **17**, 599–615.

Cliff, W. J. (1970). The aortic tunica media of aging rats. *Exp. and Mol. Path.*, **13**, 172–189.

Cliff, W. J. (1971). The ultrastructure of aortic elastica as revealed by prolonged treatment with OsO_4. *Exp. and Mol. Path.*, **15**, 220–229.

Coccheri, S. and Astrup, T. (1961). Thromboplastic and fibrinolytic activities of large human vessels. *Proc. Soc. exp. Biol. Med.*, **108**, 369–372.

Cohn, Z. A. and Wiener, E. (1963). The particulate hydrolases of macrophages. I. Comparative enzymology isolation, and properties. *J. exp. Med.*, **118**, 991–1008.

Comfort, A. (1964). *Aging: The Biology of Senescence*. Routledge and Kegan Paul, London.

Comroe, J. H., Jr (1964). Peripheral chemoreceptors. In *Handbook of Physiology, Sect. 3, Respiration*, Vol. 1, ed. W. O. Fenn and H. Rahn. American Physiological Society, Washington, pp. 557–583.

Constantinides, P. and Robinson, M. (1969a). Ultrastructural injury of arterial endothelium. II. Effects of vasoactive amines. *Archs Path.*, **88**, 106–112.

Constantinides, P. and Robinson, M. (1969b). Ultrastructural injury of arterial endothelium. I. Effects of pH, osmolarity, anoxia, and temperature. *Archs Path.*, **88**, 99–105.

Cooke, P. H. and Fay, S. F. (1972). Correlation between fiber length, ultrastructure, and the length–tension relationship of mammalian smooth muscle. *J. Cell Biol.*, **52**, 105–116.

Cookson, F. B. (1971). The origin of foam cells in atherosclerosis. *Brit. J. exp. Path.*, **52**, 62–69.

Cooper, J. H. (1969). The reducing action of elastic tissue on ferric ferricyanide. *J. Histochem. Cytochem.*, **17**, 539–544.

Copley, A. L. and Scheinthal, B. M. (1970). Nature of the endo-endothelial layer as demonstrated by ruthenium red. *Expl Cell Res.*, **59**, 491–492.

Corner, G. W. (1920). On the widespread occurrence of reticular fibrils produced by capillary endothelium. *Contrib. to Embryol.*, **9**, 87–92.

Cotran, R. S. (1965). Endothelial phagocytosis: an electron microscopic study. *Exp. and Mol. Path.*, **4**, 217–231.

Cotran, R. S. (1967). Delayed and prolonged vascular leakage in inflammation. III. Immediate delayed vascular reactions in skeletal muscle. *Exp. and Mol. Path.*, **6**, 143–155.

Cotran, R. S. and Karnovsky, M. J. (1967). Vascular leakage induced by horseradish peroxidase in the rat. *Proc. Soc. exp. Biol. Med.*, **126**, 557–561.

Cotran, R. S. and Majno, G. (1964). The delayed and prolonged vascular leakage in inflammation. *Am. J. Path.*, **45**, 261–281.

Cotran, R. S., Guttuta, Monika L. and Majno, G. (1965). Studies on inflammation. Fate of intramural vascular deposits induced by histamine. *Am. J. Path.*, **47**, 1045–1077.

Coupland, R. E. and Hopwood, D. (1966). The mechanism of the differential staining reaction for adrenaline- and noradrenaline-storing granules in tissues fixed in glutaraldehyde. *J. Anat.*, **100**, 227–243.

Courtice, F. C., Simmonds, W. J. and Steinbeck, A. W. (1951). Some investigations on lymph from a thoracic duct fistula in man. *Aust. J. exp. Biol. med. Sci.*, **29**, 201–210.

Cowdry, E. V. (1952). Aging of tissue fluids in *Cowdry's Problems of Aging*, 3rd editn, ed. A. J. Lansing. Williams and Wilkins, Baltimore. pp. 23–49.

Cox, R. C. and Little, K. (1961). An electron microscope study of elastic tissue. *Proc. Roy. Soc.*, B, **155**, 232–242.

Cox, R. C. and O'Dell, B. L. (1966). High resolution electron microscope observations on normal and pathological elastin. *J. Roy. microsc. Soc.*, **85**, 401–411.

Crane, W. A. J. and Dutta, L. P. (1963). The utilization of tritiated thymidine for deoxyribonucleic acid synthesis by the lesions of experimental hypertension in rats. *J. Path. Bact.*, **86**, 83–97.

Crawford, T. and Levene, C. I. (1953). Medial thinning in atheroma. *J. Path. Bact.*, **66**, 19–23.

Crocker, D. J., Murad, T. M. and Geer, J. C. (1970). Role of pericyte in wound healing. An ultrastructural study. *Exp. and Mol. Path.*, **13**, 51–65.

Cruikshank, B. and Hill, A. G. S. (1953). Histochemical identification of a connective tissue antigen. In *Nature and Structure of Collagen*, ed. J. T. Randall. Butterworth, London. pp. 27–32.

Curran, R. C. (1957). The elaboration of mucopolysaccharides by vascular endothelium. *J. Path. Bact.*, **74**, 347–352.

Curtis, A. S. G. (1960). Cell contacts: some physical considerations. *Am. Nat.*, **94**, 37–56.

Curtis, A. S. G. (1964). The mechanism of adhesion of cells to glass. *J. Cell Biol.*, **20**, 199–215.

Dale, H. H. (1920a). Capillary poisons and shock. *Bull. Johns Hopk. Hosp.*, **31**, 257–265.

Dale, H. H. (1920b). Anaphylaxis. *Bull. Johns Hopk. Hosp.*, **31**, 310–319.

Daoud, A., Jarmolych, J., Zumbo, A., Fani, K. and Florentin, R. (1964). 'Preatheroma' phase of coronary atherosclerosis in man. *Exp. and Mol. Path.*, **3**, 475–484.

De Bruyn, P. P. H., Breen, P. C. and Thomas, T. B. (1970). The microcirculation of the bone marrow. *Anat. Rec.*, **168**, 55–68.

Dempsey, E. W. (1952). The chemical characterization and submicroscopic structure of elastic tissue. *Science*, **116**, 520 (Abstract).

Dempsey, E. W. and Lansing, A. I. (1954). Elastic tissue. *Int. Rev. Cytol.*, **3**, 437–453.

De Petris, S., Karlsbad, G. and Pernis, B. (1962). Filamentous structures in the cytoplasm of normal mononuclear phagocytes. *J. Ultrastruct. Res.*, **7**, 39–55.

Devine, C. E. and Somlyo, A. P. (1971). Thick filaments in vascular smooth muscle. *J. Cell Biol.*, **49**, 636–649.

Devis, R. and James, D. W. (1962). Electron microscopic appearance of close relationships between adult guinea pig fibroblasts in tissue culture. *Nature*, **194**, 695–696.

Dewey, M. M. and Barr, L. (1962). Intercellular connection between smooth muscle cells: the nexus. *Science*, **137**, 670.

Dickinson, C. J. and Secker Walker, R. H. (1970). Longitudinal distribution of blood in the rabbit in relation to the heart, with observations on the contribution of different organs. *Circulation Res.*, **27**, 851–861.

Dintenfass, L. (1967). Inversion of the Fahraeus–Lindqvist phenomenon in blood flow through capillaries of diminishing radius. *Nature*, **215**, 1099–1100.

Dobb, M. G. (1964). Protofibrils in α-keratin. *J. molec. Biol.*, **10**, 156.

Dobb, M. G. (1966). The structure of keratin protofibrils. *J. Ultrastruct. Res.*, **14**, 294–299.

Dobb, M. G., Fraser, R. D. B. and Macrae, T. P. (1967). The fine structure of silk fibroin. *J. Cell Biol.*, **32**, 289–295.

Duchacek, H. (1967). Unsteady flow in tubes. In *Microvascular Surgery*, ed. R. M. P. Donaghy and M. G. Yaşargil. Georg Thieme Verlag, Stuttgart. pp. 15–19.

Duff, G. L. (1932). Vital staining of the rabbit's aorta in the study of arteriosclerosis. *Am. J. Path.*, **8**, 219–234.

Duguid, J. B. (1954). Diet and coronary disease. *Lancet*, **1**, 891–895.

Duling, B. R. and Berne, R. M. (1971). Oxygen and the local regulation of blood flow: possible significance of longitudinal gradients in arterial blood oxygen tension. *Circulation Res.*, **28**, Suppl. I, 65–69.

Dunihue, F. H. (1967). Histology of small blood vessels. In *Microvascular Surgery*, ed. R. M. P. Donaghy and M. G. Yaşargil. Georg Thieme Verlag, Stuttgart. pp. 14–15.

Dustin, A. P. and Chodkowski, K. (1938). Étude de la cicatrisation par la réaction colchicinique. *Arch Int de Méd Exp.*, **13**, 641–662.

Duve, C. de (1959). Lysosomes, a new group of cytoplasmic particles. In *Subcellular Particles*, ed. T. Hayashi. Ronald Press Co., N.Y. pp. 128–159.

Easty, G. C. and Mercer, E. H. (1962). An electron microscope study of model tissues formed by the agglutination of erythrocytes. *Expl Cell Res.*, **28**, 215–227.

Ebert, R. H. and Florey, H. W. (1939). The extravascular development of the monocyte observed *in vivo*. *Brit. J. exp. Path.*, **20**, 342–356.

Edwards, L. C. and Dunphy, J. E. (1958). Wound healing. *New Engl. J. Med.*, **259** (2 parts), 224–232 and 275–284.

Edwards, R. H., Sarmenta, S. S. and Hass, G. M. (1960). Stimulation of granulation tissue growth by tissue extracts. *Archs Path.*, **69**, 286–302.

Ehinger, B., Falck, B. and Sporrong, B. (1966). Adrenergic fibres to the heart and to peripheral vessels. *Bibl. Anat.*, **8**, 35–45.

Elliott, D. F., Horton, E. W. and Lewis, G. P. (1960). Actions of pure bradykinin. *J. Physiol.*, **153**, 473–480.

Ellis, H. D. (1970). Effects of shear treatment on drag-reducing polymer solutions and fibre suspensions. *Nature*, **226**, 352–353.

Ellison, G. D. and Zanchetti, A. (1971). Specific appearance of sympathetic cholinergic vasodilatation in muscles during conditioned movements. *Nature*, **232**, 124–125.

Elsner, R., Hanafee, W. N. and Hammond, D. D. (1971). Angiography of the inferior vena cava of the harbor seal during simulated diving. *Am. J. Physiol.*, **220**, 1155–1157.

Esterly, J. R. and Oppenheimer, E. H. (1967). Vascular lesions in infants with congenital Rubella. *Circulation*, **37**, 544–554.

Estes, P. C. and Cheville, N. F. (1970). The ultrastructure of vascular lesions in equine viral arteritis. *Am. J. Path.*, **58**, 235–253.

Evans, C. L. (1918–19). The velocity factor in cardiac work. *J. Physiol.*, **52**, 6–14.

Evans, C. L. (1949). *Principles of Human Physiology*, 10th editn. Churchill, London.

Fahrenbach, W. H., Sandberg, L. B. and Cleary, E. G. (1966). Ultrastructural studies on early elastogenesis. *Anat. Rec.*, **155**, 563–575.

Falck, B. (1962). Observations on the possibilities of the cellular localization of monoamines by a fluorescence method. *Acta physiol. scand.*, **56**, Suppl. 197.

Falck, B. (1964). Cellular localization of monoamines. *Progress in Brain Research*, **8**, 28–44.

Fanger, H. and Barker, B. E. (1960). Capillaries of normal and diseased breast. *Archs Path.*, **69**, 67–76.

Farnes, P. and Barker, B. E. (1963). Cytochemical studies of human bone marrow fibroblast-like cells. I. Alkaline phosphatase. *Expl Cell Res.*, **29**, 278–288.

Farquhar, M. G. (1960). An electron microscope study of glomerular permeability. *Anat. Rec.*, **136**, 191 (Abstract).

Farquhar, M. G. and Palade, G. E. (1960). Segregation of ferritin in glomerular protein absorption droplets. *J. biophys. biochem. Cytol.*, **7**, 297–304.

Farquhar, M. G. and Palade, G. E. (1962). Functional evidence for the existence of a third cell type in the renal glomerulus. *J. Cell Biol.*, **13**, 55–87.

Farquhar, M. G. and Palade, G. E. (1963). Junctional complexes in various epithelia. *J. Cell Biol.*, **17**, 375–412.

Farrant, J. L. (1954). An electron microscopic study of ferritin. *Biochim. biophys. Acta*, **13**, 569–576.

Fawcett, D. W. (1959). The fine structure of capillaries, arterioles and small arteries. In *The Microcirculation*, Symposium on factors influencing exchange of substances across capillary wall, ed. S. R. M. Reynolds and B. W. Zweifach. University of Illinois Press. pp. 1–27.

Fawcett, D. W. (1963). Comparative observations on the fine structure of blood capillaries. In *The Peripheral Blood Vessels*, ed. J. L. Orbison and D. E. Smith. Williams and Wilkins, Baltimore. pp. 17–44.

Fedorko, M. E. and Hirsch, J. G. (1971). Studies on transport of macromolecules and small particles across mesothelial cells of the mouse omentum. I. Morphologic aspects. *Expl Cell Res.*, **69**, 113–127.

Feldman, S. A. and Glagov, S. (1971). Transmedial collagen and elastin gradients in human aortas: reversal with age. *Atherosclerosis*, **13**, 385–394.

Felix, M. D. and Dalton, A. J. (1956). A comparison of mesothelial cells and macrophages in mice after the intraperitoneal inoculation of melanin granules. *J. biophys. biochem. Cytol.*, **2**, Suppl., 109–114.

Ferrer, J. (1957). Correlation of the vascular supply of the pars distalis of the rat's adenohypophysis with the distribution of the basophilic cells. *Anat. Rec.*, **127**, 527–537.

Fillenz, M. (1970). Innervation of pulmonary and bronchial blood vessels of the dog. *J. Anat.*, **106**, 449–461.

First Conference on Microcirculatory Physiology and Pathology (1954). *Anat. Rec.*, **120**, 241–361.

Florey, Lord (1966). The endothelial cell. *Brit. med. J.*, **2**, 487–490.

Florey, H. W., Poole, J. C. F. and Meek, G. A. (1959). Endothelial cells and 'cement' lines. *J. Path. Bact.*, **77**, 625–636.

Florey, H. W., Greer, S. J., Kiser, J., Poole, J. C. F., Telander, R. and Werthessen, N. T. (1962). The development of the pseudointima lining fabric grafts of the aorta. *Brit. J. exp. Path.*, **43**, 655–660.

Folkow, B., Haglund, U., Jodal, M. and Lundgren, O. (1971). Blood flow in the calf muscle of man during heavy rhythmic exercise. *Acta physiol. scand.*, **81**, 157–163.

Folkow, B., Hallbäck, M., Lundgren, Y. and Weiss, L. (1970). Background of increased flow resistance and vascular 'reactivity' in spontaneously hypertensive rats. *Acta physiol. scand.*, **79**, 42A–43A.

Fontaine, R., Ebel, A., Pantesco, V., Kempf, E. and Mack, G. (1968). Contribution à l'étude du compartement biochimique de la paroi artérielle au cours du vieillissement et de l'atherogenase. Similitudes et différences dans ces deux états. *Angiology*, **19**, 172–197.

Foot, N. C. (1921). Studies on endothelial reactions. V. The endothelium in the healing of aseptic wounds in the omentum of rabbits. *J. exp. Med.*, **34**, 625–642.

Foreman, J. E. K. and Hutchison, K. J. (1970). Arterial wall vibration distal to stenoses in isolated arteries of dog and man. *Circulation Res.*, **26**, 583–590.

Fowler, N. O., Westcott, R. N. and Scott, R. C. (1953). Normal pressure in the right heart and pulmonary artery. *Am. Heart J.*, **46**, 264–267.

Franchi, C. M. and De Robertis, E. (1951). Electron microscope observations on elastic fibers. *Proc. Soc. exp. Biol. Med.*, **76**, 515–518.

Franke, W. W. (1970). Attachment of muscle filaments to the outer membrane of the nuclear envelope. *Z. Zellforsch. mikrosk. Anat.*, **111**, 143–148.

Franklin, K. J. (1927–28). Valves in veins: an historical survey. *Proc. Roy. Soc. Med.*, **21**, 1–33.

Franklin, K. J. (1928). The physiology and pharmacology of veins. *Physiol. Rev.*, **8**, 346–364.

Franklin, K. J. (1949). A short history of physiology. 2nd editn. Staples, London and N.Y.

Franklin, K. J. and Haynes, F. (1927). The histology of the giraffe's carotid, functionally considered. *J. Anat.*, **62**, 115–117.

French, J. E. (1963). Endothelial structure and function. In *Evolution of the Atherosclerotic Plaque*, ed. R. J. Jones. University of Chicago Press, Chicago. pp. 15–28.

French, J. E. (1970). The structure and function of the blood vessel wall. In *Vascular Factors and Thrombosis*, ed. K. M. Brinkhous *et al.* Schattauer Verlag, Stuttgart and N.Y. pp. 11–22.

French, J. E., Jennings, M. A., Poole, J. C. F., Robinson, D. S. and Sir Howard Florey (1963). Intimal changes in the arteries of aging swine. *Proc. Roy. Soc.*, B, **158**, 24–42.

Friedman, M., Byers, S. O. and St George, S. (1966). Site of origin of the luminal foam cells of atherosclerosis. *Am. J. clin. Path.*, **45**, 238–246.

Fritz, K. E., Jarmolych, J. and Daoud, A. S. (1970). Association of DNA synthesis and apparent dedifferentiation of aortic smooth muscle cells *in vitro*. *Exp. and Mol. Path.*, **12**, 354–362.

Furchgott, R. F. and Bhadrakom, S. (1958). Reaction of strips of rabbit aorta to epinephrine, isopropylarterenol, sodium nitrite and other drugs. *J. Pharmac. exp. Ther.*, **108**, 129–143.

Furness, J. B. and Moore, J. (1970). The adrenergic innervation of the cardiovascular system of the lizard *Trachysaurus rugosus*. *Z. Zellforsch. mikrosk. Anat.*, **108**, 150–176.

Furnival, C. M., Linden, R. J. and Snow, H. M. (1970). The effect of hypoxia on the pulmonary veins. *J. Physiol.*, **210**, 43–44P.

Gabella, G. (1971). Relationship between sarcoplasmic reticulum and caveolae intracellulares in the intestinal smooth muscle. *J. Physiol.*, **216**, 42–44P.

Garamvölgyi, N., Vizi, E. S. and Knoll, J. (1971). The regular occurrence of thick filaments in stretched mammalian smooth muscle. *J. Ultrastruct. Res.*, **34**, 135–143.

Garbarsch, C., Matthiessen, M. E., Helin, P. and Lorenzen, I. (1969). Arteriosclerosis and hypoxia. I. Gross and microscopic changes in rabbit aorta induced by systemic hypoxia. Histochemical studies. *J. Atheroscler. Res.*, **9**, 283–294.

Gautvik, K. (1970). Parasympathetic neuro-effector transmission and functional vasodilatation in the submandibular salivary glands of cats. *Acta physiol. scand.*, **79**, 204–215.

Geer, J. C., McGill, H. C., Jr and Strong, J. P. (1961). The fine structure of human atherosclerotic lesions. *Am. J. Path.*, **38**, 263–275.

Geiringer, E. (1951). Intimal vascularisation and atherosclerosis. *J. Path. Bact.*, **63**, 201–211.

Gerrity, R. G. (1972). Post-natal development of the rat aorta: a morphological and biochemical analysis. Thesis submitted for the degree of Doctor of Philosophy, Australian National University.

Gerrity, R. G. and Cliff, W. J. (1972). The aortic tunica intima in young and aging rats. *Exp. and Mol. Path.*, **16**, 382–402.

References

Gersh, I. and Catchpole, H. R. (1949). The organisation of ground substance and basement membrane and its significance in tissue injury, disease and growth. *Am. J. Anat.*, **85**, 457–522.

Ghidoni, J. J. and O'Neal, R. M. (1967). Recent advances in molecular pathology: a review. Ultrastructure of human atheroma. *Exp. and Mol. Path.*, **7**, 378–400.

Giacomelli, F., Wiener, J. and Spiro, D. (1970). The cellular pathology of experimental hypertension. Increased permeability of cerebral arterial vessels. *Am. J. Path.*, **59**, 133–159.

Gillman, T. (1959). Reduplication, remodeling, regeneration, repair, and degeneration of arterial elastic membranes. Some implications for the pathogenesis of arterial diseases. *Archs Path.*, **67**, 624–642.

Gillman, T., Hathorn, M. and Penn, J. (1957). Micro-anatomy and reactions to injury of vascular elastic membranes and associated polysaccharides. In *Connective Tissue, a Symposium*, ed. R. E. Tunbridge. Blackwell, Oxford. pp. 120–135.

Gimbrone, M. A., Jr, Cotran, R. S. and Folkman, J. (1974). Human vascular endothelial cells in culture, growth and DNA synthesis. *J. Cell Biol.*, **60**, 673–684.

Glagov, S. (1965). Hemodynamic factors in localisation of atherosclerosis. *Acta Cardiol.* Suppl. 11, 311–337.

Glagov, S. and Wolinsky, H. (1963). Aortic wall as a 'two-phase' material. *Nature*, **199**, 606–608.

Glasser, O. (ed.) (1944). *Medical Physics*, Vol. 1. See Chapter by H. N. Holmes: Colloid chemistry. Year Book Publishers Inc., Chicago. pp. 257–264.

Glasser, O. (ed.) (1950). *Medical Physics*, Vol. 2. See Chapter by E. N. Harvey: Bubble formation in liquids. Year Book Publishers Inc., Chicago. pp. 137–150.

Goldfischer, S., Novikoff, A. B., Albala, A. and Biempica, L. (1970). Hemoglobin uptake by rat hepatocytes and its breakdown within lysosomes. *J. Cell Biol.*, **44**, 513–529.

Goldsmith, H. L. (1970). Motion of particles in a flowing system. In *Vascular Factors and Thrombosis*, ed. K. M. Brinkhous *et al*. Schattauer Verlag, Stuttgart and N.Y. pp. 91–110.

Goodenough, D. A. and Revel, J. P. (1970). A fine structural analysis of intercellular junctions in the mouse liver. *J. Cell Biol.*, **45**, 272–290.

Goodman, P., Latta, J. S., Wilson, R. B. and Kadis, B. (1968). The fine structure of sow lutein cells. *Anat. Rec.*, **161**, 77–90.

Gore, I., Fujinami, T. and Shirahama, T. (1965). Endothelial changes produced by ascorbic acid deficiency in guinea pigs. *Archs Path.*, **80**, 371–376.

Gore, I., Takada, M. and Austin, J. (1970). Ultrastructural basis of experimental thrombocytopenic purpura. *Archs Path.*, **90**, 197–205.

Gotte, L. and Serafini-Fracassini, A. (1963). Electron microscope observations on the structure of elastin. *J. Atheroscler. Res.*, **3**, 247–251.

Gotte, L., Meneghelli, V. and Castellani, A. (1965). Electron microscope observations and chemical analyses of human elastin. In *Structure and Function of Connective and Skeletal Tissue*, proceedings of an Advanced Study Institute organised under the auspices of NATO, St Andrews. Butterworth, London. pp. 93–119.

Gowans, J. L. and Knight, E. J. (1964). The route of re-circulation of lymphocytes in the rat. *Proc. Roy. Soc.*, B, **159**, 257–282.

Grafflin, A. L. and Bagley, E. H. (1953). Studies of peripheral blood vascular beds. *Bull. Johns Hopk. Hosp.*, **92**, 47–73.

Graham, R. C., Jr and Karnovsky, M. J. (1966). Glomerular permeability ultrastructural cytochemical studies using peroxidases as protein tracers. *J. exp. Med.*, **124**, 1123–1133.

Graham, R. C., Ebert, R. N., Ratnoff, O. D. and Moses, J. M. (1965). Pathogenesis of inflammation. II. *In vivo* observations of the inflammatory effects of activated Hageman factor and bradykinin. *J. exp. Med.*, **121**, 807–818.

Grant, L., Palmer, P. and Sanders, A. G. (1962). The effect of heparin on the sticking of white cells to endothelium in inflammation. *J. Path. Bact.*, **83**, 127–133.

Grant, R. A. (1967). Content and distribution of aortic collagen, elastin and carbohydrate in different species. *J. Atheroscler. Res.*, **7**, 463–472.

Grant, R. A., Horne, R. W. and Cox, R. W. (1965). New model for the tropocollagen macromolecule and its mode of aggregation. *Nature*, **207**, 822–826.

Grant, R. T. (1966). The effects of denervation on skeletal muscle blood vessels (rat cremaster). *J. Anat.*, **100**, 305–316.

Grant, R. T. and Payling Wright, H., (1971). The peculiar vasculature of the external spermatic fascia in the rat; possibly subserving thermoregulation. *J. Anat.*, **109**, 293–305.

Grant, R. T., Bland, E. F. and Camp, P. D. (1932). Observations on the vessels and nerves of the rabbit's ear with special reference to the reaction to cold. *Heart*, **16**, 69–101.

Gray, S. H., Handler, F. P., Blache, J. O., Zuckner, J. and Blumenthal, H. T. (1953). Aging processes of aorta and pulmonary artery in Negro and White races. Comparative study of various segments. *Archs Path.*, **56**, 238–253.

Green, H. D. (1950). Circulatory system: physical principles. In *Medical Physics*, Vol. 2, ed. O. Glasser. The Year Book Publishers Inc., Chicago. pp. 228–251.

Green, H. D., Lewis, R. N., Nickerson, N. D. and Heller, A. L. (1944). Blood flow, peripheral resistance and vascular tonus, with observations on the relationship between blood flow and cutaneous temperature. *Am. J. Physiol.*, **141**, 518–536.

Greenlee, T. K., Ross, R. and Hartman, J. L. (1966). The fine structure of elastic fibers. *J. Cell Biol.*, **30**, 59–71.

Greenway, C. V. and Stark, R. D. (1971). Hepatic vascular bed. *Physiol. Rev.*, **51**, 23–65.

Gregg, D. E. (1934). The plasic variations in coronary flow, studied by an autoperfusion method. *Am. J. Physiol.*, **109**, 44 (Abstract).

Gresham, G. A., Howard, A. N. and King, A. J. (1962). A comparative histopathological study of the early atherosclerotic lesion. *Brit. J. exp. Path.*, **43**, 21–23.

Groat, R. A. (1948). Relationship of volumetric rate of blood flow to arterial diameter. *Fedn. Proc. Fedn. Am. Socs. exp. Biol.*, **7**, 45 (Abstract).

Gross, J. (1949). The structure of elastic tissue as studied with the electron microscope. *J. exp. Med.*, **89**, 699–708.

Grotte, G. (1956). Passage of dextran molecules across the blood–lymph barrier. *Acta chir. scand.*, Suppl. 211, 1–84.

Grubb, D. J. and Jones, A. L. (1971). Ultrastructure of hepatic sinusoids in sheep. *Anat. Rec.*, **170**, 75–80.

Guntheroth, W. G. and Chakmakjian, S. (1971). Active changes in tone in the canine vena cava. *Circulation Res.*, **28**, 554–558.

Gutstein, W. H., Farrell, G. A. and Schenck, D. J. (1970). *In vivo* demonstration of junctional blood flow disturbance by hot wire anemometry. *Atherosclerosis*, **11**, 485–496.

Haar, J. L. and Ackerman, G. A. (1971). A phase and electron microscopic study of vasculogenesis and erythropoiesis in the yolk sac of the mouse. *Anat. Rec.*, **170**, 199–224.

Hadfield, G. (1951). Granulation tissue. *Ann. Roy. Coll. Surg.*, **9**, 397–407.

Haimovici, H., Maier, N. and Strauss, L. (1958). Fate of aortic homografts in experimental canine atherosclerosis: study of fresh thoracic implants into abdominal aorta. *Archs Surg.*, **76**, 282–288.

Hall, C. E. and Doty, P. (1958). A comparison between dimensions of some macromolecules determined by electron microscopy and by physical chemical methods. *J. Am. chem. Soc.*, **80**, 1269–1274.

Hall, D. A. (1961). The possible implications of enzymes of the elastase complex in atherosclerosis. *J. Atheroscler. Res.*, **1**, 173–183.

Hall, D. A., Reed, R. and Tunbridge, R. E. (1955). Electron microscope studies of elastic tissue. *Expl Cell Res.*, **8**, 35–48.

Hallbäck, M., Lundgren, Y. and Weiss, L. (1971). Reactivity to noradrenaline of aortic strips and portal veins from spontaneously hypertensive and normotensive rats. *Acta physiol. scand.*, **81**, 176–181.

Hallock, P. and Benson, I. C. (1937). Studies on the elastic properties of human isolated aorta. *J. clin. Invest.*, **16**, 595–602.

Hama, K. (1960). The fine structure of some blood vessels of the earthworm. *J. biophys. biochem. Cytol.*, **7**, 717–724.

Hama, K. (1961). On the existence of filamentous structures in endothelial cells of the amphibian capillary. *Anat. Rec.*, **139**, 437–441.

Hammond, G. L. and Moggio, R. A. (1971). Function of microvascular pathways in coronary circulation. *Am. J. Physiol.*, **220**, 1463–1467.

Harker, L. A. and Slichter, S. J. (1970). Studies of platelet and fibrinogen kinetics in patients with prosthetic heart valves. *New Engl. J. Med.*, **283**, 1302–1305.

Harris, H. (1954). Role of chemotaxis in inflammation. *Physiol. Rev.*, **34**, 529–562.

Harris, H. (1960). Mobilization of defensive cells in inflammatory tissue. *Bact. Rev.*, **24**, 3–15.

Hass, G. M. (1939a). Elastic tissue (Review). *Archs Path.*, **27**, 334–365.

Hass, G. M. (1939b). Elastic tissue (Review). *Archs Path.*, **27**, 583–613.

Hass, G. M. (1943). Elastic tissue. III. Relations between the structure of the aging aorta and the properties of the isolated aortic elastic tissue. *Archs Path.*, **35**, 29–45.

Hass, G. M. (1955). Observations on vascular structure in relation to human and experimental arteriosclerosis. In *Symposium on Atherosclerosis*. Publ. No. 338. Nat. Acad. Sci., National Research Council. pp. 24–32.

Hass, G. M. (1963). Metabolic and nutritional factors in peripheral vascular disease. In *The Peripheral Blood Vessels*, ed. J. L. Orbison and D. E. Smith. Williams and Wilkins, Baltimore. pp. 157–204.

Hassler, O. (1962). Physiological intima cushions in the large cerebral arteries of young individuals. I. Morphological structure and possible significance for the circulation. *Acta path. microbiol. scand.*, **55**, 19–27.

Hassler, O. (1969). A senile vascular change resulting from excessive spiralling of arteries. *J. Geront.*, **24**, 37–41.

Hassler, O. (1970). The origin of the cells constituting arterial intima thickening, an experimental autoradiographic study with the use of H^3-thymidine. *Lab. Invest.*, **22**, 286–293.

Haust, M. D. (1965). Fine fibrils of extracellular space (microfibrils): their structure and role in connective tissue organization. *Am. J. Path.*, **47**, 1113–1137.

Haust, M. D., Wyllie, J. C. and More, R. H. (1965). Electron microscopy of fibrin in human atherosclerotic lesions. *Exp. and Mol. Path.*, **4**, 205–216.

Haust, M. D., More, R. H., Bencosme, S. A. and Balis, J. V. (1965). Elastogenesis in human aorta: an electron microscopic study. *Exp. and Mol. Path.*, **4**, 508–524.

Heath, D., Wood, E. H., Du Shane, J. W. and Edwards, J. E. (1959). The structure of the pulmonary trunk at different ages and in cases of pulmonary hypertension and pulmonary stenosis. *J. Path. Bact.*, **77**, 443–456.

Helfer, H. and Jaques, R. (1970). A method for evaluating tone, frequency and venous output of the isolated murine portal vein. The effect of adrenergic and cholinergic stimulants. *Pharmacology*, **4**, 65–79.

Heron, I. (1971). The transplanted rabbit heart. Histological, immunofluorescent and electrocardiographic changes. *Acta path. microbiol. scand., A*, **79**, 373–380.

Higginbotham, A. C., Higginbotham, F. H. and Williams, T. W. (1963). Vascularization of blood vessel walls. In *Evolution of the Atherosclerotic Plaque*, ed. R. J. Jones. University of Chicago Press, Chicago. pp. 265–277.

Hill, C. H. (1969). A role of copper in elastin formation. *Nutr. Abstr. Rev.*, **27**, 99–100.

Hilton, S. M. (1962). Local mechanisms regulating peripheral blood flow. *Physiol. Rev.*, **42**, Suppl. 5, 265–275.

Hodge, A. J. (1956). Effects of the physical environment on some lipoprotein layer systems and observations on their morphogenesis. *J. biophys. biochem. Cytol.*, **2**, Suppl., 221–227.

Hoff, H. F. and Graf, J. (1966). An electron microscopy study of phosphatase activity in the endothelial cells of rabbit aorta. *J. Histochem. Cytochem.*, **14**, 719–724.

Hogan, M. J. and Feeney, L. (1963). The ultrastructure of the retinal blood vessels. I. The large vessels. *J. Ultrastruct. Res.*, **9**, 10–28.

Holfreter, J. (1947). Changes of structure and the kinetics of differentiating embryonic cells. *J. Morph.*, **80**, 57–89.

Hollander, W. (1967). Recent advances in experimental and molecular pathology. Influx, synthesis, and transport of arterial lipoproteins in atherosclerosis. *Exp. and Mol. Path.*, **7**, 248–258.

Horowitz, R. E., Burrows, L., Paronetto, F., Dreiling, D. and Kark, A. F. (1965). Immunologic observations on homografts. II. The canine kidney. *Transplantation*, **3**, 318–325.

Howard, J. G., Boak, J. L. and Christie, G. H. (1966). Further studies on the transformation of thoracic duct cells into liver macrophages. *Ann. N.Y. Acad. Sci.*, **129**, 327–339.

Hughes, A. F. W. and Dann, L. (1941). Vascular regeneration in experimental wounds and burns. *Brit. J. exp. Path.*, **22**, 9–14.

Hume, D. M. (1968–9). *The Immunological Consequences of Organ Homotransplantation in Man*. Harvey Lect., New York, Series 64, 261–388.

Hunter, D. (1956). (ed.) *Price's Textbook of the Practice of Medicine*. Oxford University Press, London.

Hurley, J. V. (1963). Incubation of serum with tissue extracts as a cause of chemotaxis of granulocytes. *Nature*, **198**, 1212–1213.

Hurley, J. V. (1964). Acute inflammation: the effect of concurrent leucocytic emigration and increased permeability on particle retention by the vascular wall. *Brit. J. exp. Path.*, **45**, 627–633.

Hurley, J. V., Ham, K. N. and Ryan, G. B. (1967). The mechanism of the delayed prolonged phase of increased vascular permeability in mild thermal injury in the rat. *J. Path. Bact.*, **94**, 1–12.

Hüttner, I., More, R. H. and Rona, G. (1970). Fine structural evidence of specific mechanism for increased endothelial permeability in experimental hypertension. *Am. J. Path.*, **61**, 395–404.

Huxley, H. E. (1957). The double array of filaments in cross-striated muscle. *J. biophys. biochem. Cytol.*, **3**, 631–648.

Illingworth, C. F. W. (1955). *A Short Textbook of Surgery*, 6th editn. Churchill, London.

Intaglietta, M. and Zweifach, B. W. (1971). Geometrical model of the microvasculature of rabbit omentum from *in vivo* measurements. *Circulation Res.*, **28**, 593–600.

Irey, N. S., Manion, W. C. and Taylor, H. B. (1970). Vascular lesions in women taking oral contraceptives. *Archs Path.*, **89**, 1–8.

Jackson, D. S. (1957). Connective tissue growth stimulated by carrageenin. *Biochem. J.*, **65**, 277–284.

Jackson, D. S. and Cleary, E. G. (1967). The determination of collagen and elastin. In *Methods of Biochemical Analysis*, Vol. 15, ed. D. Glick. Interscience, N.Y. pp. 25–76.

Jackson, S. F. (1956). The morphogenesis of avian tendon. *Proc. Roy. Soc.*, B, **144**, 556–572.

Jacobson, W. (1953). Histological survey of the normal connective tissue and its derivations. In *Nature and Structure of Collagen*, ed. J. T. Randall. Butterworth, London. pp. 6–13.

Jellinek, H., Veress, B., Bálint, A. and Nagy, Z. (1970). Lymph vessels of rat aorta and their changes in experimental atherosclerosis: an electron microscopic study. *Exp. and Mol. Path.*, **13**, 370–376.

Jennings, M. A. and Lord Florey (1967). An investigation of some properties of endothelium related to capillary permeability. *Proc. Roy. Soc.*, B, **167**, 39–63.

Jennings, M. A., Marchesi, V. T. and Florey, H. W. (1962). The transport of particles across the walls of small blood vessels. *Proc. Roy. Soc.*, B, **156**, 14–19.

Jennings, M. A., Brock, L. G. and Lord Florey (1966). A comparison of connective tissue lining aortic grafts with extravascular connective tissue. *Proc. Roy. Soc.*, B, **165**, 206–223.

Johansen, K. and Martin, A. W. (1965). Comparative aspects of cardiovascular function in vertebrates. In *Handbook of Physiology*, Sect. 2, Circulation, Vol. 3, ed. W. F. Hamilton and P. Dow. Am. Physiol. Soc., Washington. pp. 2583–2614.

Johansson, B. and Bohr, D. F. (1966). Rhythmic activity in smooth muscle from small subcutaneous arteries. *Am. J. Physiol.*, **210**, 801–806.

Johnson, R. A. (1969). Lymphatics of blood vessels. *Lymphology*, **2**, 44–56.

Johnson, R. A. and Blake, T. M. (1965). Lymphatics of arteries. *Circulation*, **32**, Suppl. ii, 119.

Johnson, S. A., Balboa, R. S., Dessel, B. H., Monto, R. W., Siegesmund, K. A. and Greenwalt, T. J. (1964). The mechanism of the endothelial supporting function of intact platelets. *Exp. and Mol. Path.*, **3**, 115–127.

Joó, F. (1971). Increased production of coated vesicles in the brain capillaries during enhanced permeability of the blood–brain barrier. *Brit. J. exp. Path.*, **52**, 646–649.

Jordan, G. L., Jr, Stump, M. M., De Bakey, M. E. and Halpert, B. (1962). Endothelial lining of Dacron prostheses of porcine thoracic aortas. *Proc. Soc. exp. Biol. Med.*, **110**, 340–343.

Kabat, E. A. and Furth, J. (1941). A histochemical study of the distribution of alkaline phosphatase in various normal and neoplastic tissues. *Am. J. Path.*, **17**, 303–318.

Kampmeier, O. F. and Birch, C. L. (1927). The origin and development of the venous valves with particular reference to saphenous district. *Am. J. Anat.*, **38**, 451–499.

Kao, K.-Y. T. and McGavack, T. H. (1959). Connective tissue. I. Age and sex influence on protein composition of rat tissues. *Proc. Soc. exp. Biol. Med.*, **101**, 153–157.

Kaplan, D. and Meyer, K. (1960). Mucopolysaccharides of aorta at various ages. *Proc. Soc. exp. Biol. Med.*, **105**, 78–81.

Karatzas, N. B., Noble, M. I. M., Saunders, K. B. and McIlroy, M. B. (1970). Transmission of the blood flow pulse through the pulmonary arterial tree of the dog. *Circulation Res.*, **27**, 1–9.

Karnovsky, M. J. (1967). The ultrastructural basis of capillary permeability studied with peroxidase as a tracer. *J. Cell Biol.*, **35**, 213–236.

Karnovsky, M. J. (1968). The ultrastructural basis of transcapillary exchanges. *J. gen. Physiol.*, **52**, Suppl., 64–95.

Karpinski, E., Barton, S. and Schachter, M. (1971). Vasodilator nerve fibres to the submaxillary gland of the cat. *Nature*, **232**, 122–124.

Karrer, H. E. (1960a). The striated musculature of blood vessels. II. Cell interconnections and cell surface. *J. biophys. biochem. Cytol.*, **8**, 135–150.

Karrer, H. E. (1960b). Electron microscopy study of developing chick embryo aorta. *J. Ultrastruct. Res.*, **4**, 420–454.

Karrer, H. E. (1960c). Electron microscopic study of the phagocytosis process in lung. *J. biophys. biochem. Cytol.*, **7**, 357–365.

Karrer, H. E. (1961). An electron microscope study of the aorta in young and in aging mice. *J. Ultrastruct. Res.*, **5**, 1–27.

Kasai, T. and Pollak, O. J. (1964). Smooth muscle cells in aortic cultures of untreated and cholesterol-fed rabbits. *Z. Zellforsch. mikrosk. Anat.*, **62**, 743–752.

Kasai, T., Pollak, O. J. and Nagasawa, S. (1964). Endothelial cells and macrophages in aortic cultures of untreated and cholesterol-fed rabbits. *Z. Zellforsch. mikrosk. Anat.*, **62**, 753–761.

Keech, M. K. (1960a). Electron microscope study of the normal rat aorta. *J. biophys. biochem. Cytol.*, **7**, 533–538.

Keech, M. K. (1960b). Electron microscope study of the lathyritic rat aorta. *J. biophys. biochem. Cytol.*, **7**, 539–545.

Kelly, F. B., Jr, Taylor, C. B. and Hass, G. M. (1952). Experimental atheroarteriosclerosis localization of lipids in experimental arterial lesions of rabbits with hypercholesterolaemia. *Archs Path.*, **53**, 419–436.

Kelly, L. S., Brown, B. A. and Dobson, E. L. (1962). Cell division and phagocytic activity in liver reticulo-endothelial cells. *Proc. Soc. exp. Biol. Med.*, **110**, 555–559.

Kelly, R. E. and Rice, R. V. (1968). Localization of myosin filaments in smooth muscle. *J. Cell Biol.*, **37**, 105–116.

Keyserlingk, D. Graf. (1970). Ultrastruktur glycerinextrahierter Dünndarmmuskelzellen der Ratte vor und nach Kontraktion. *Z. mikrosk.-anat. Forsch.*, **111**, 559–571.

King, A. L. and Lawton, R. W. (1950). Elasticity of body tissues. In *Medical Physics*, Vol. 2, ed. O. Glasser. The Year Book Publishers Inc., Chicago. pp. 303–316.

Kirk, J. E. and Laursen, T. J. S. (1955). Diffusion coefficients of various solutes for human aortic tissue. With special reference to variation in tissue permeability with age. *J. Geront.*, **10**, 288–302.

Kisch, B. (1957). Electron microscopy of the capillary wall. II. Filiform processes of the endothelium. *Expl Med. Surg.*, **15**, 89–99.

Knieriem, H. J., Kao, V. C. Y. and Wissler, R. W. (1967). Actomyosin and myosin and the deposition of lipids and serum lipoproteins. *Archs Path.*, **84**, 118–129.

Knieriem, H. J., Kao, V. C. Y. and Wissler, R. W. (1968). Demonstration of smooth muscle cells in bovine arteriosclerosis. *J. Atheroscler. Res.*, **8**, 125–136.

Korner, P. I. (1971). Integrative neural cardiovascular control. *Physiol. Rev.*, **51**, 312–367.

Kraemer, D. M. and Miller, H. (1953). Elastin content of the albuminoid fraction of human aortae. *Archs Path.*, **55**, 70–72.

Kramer, H. and Little, K. (1953). Nature of reticulin. In *Nature and Structure of Collagen*, ed. J. T. Randall. Butterworth, London. pp. 33–43.

Krogh, A. (1959). *The Anatomy and Physiology of Capillaries*. Hafner Publishing Co., N.Y.

References

Kurtz, S. M. and Feldman, J. D. (1962). Experimental studies on the formation of the glomerular basement membrane. *J. Ultrastruct. Res.*, **6**, 19–27.

Kuwabara, T. and Cogan, D. G. (1963). Retinal vascular patterns. VI. Mural cells of the retinal capillaries. *Archs Ophthal. N.Y.*, **69**, 492–502.

Ladányi, P. and Lelkes, Gy. (1968). Study of experimental fibrillogenesis in tunica muscularis of the rat ureter. *Acta Morphologica Acad. Sci. Hung.*, **16**, 147–156.

Landis, E. M. (1926). Capillary pressure in frog mesentery as determined by micro-injection methods. *Am. J. Physiol.*, **75**, 548–570.

Landis, E. M. (1927). Micro-injection studies of capillary permeability. I. Factors in the production of capillary stasis. *Am. J. Physiol.*, **81**, 124–142.

Landis, E. M. (1929). Micro-injection studies of capillary blood pressure in human skin. *Heart*, **15**, 209–228.

Landis, E. M. (1964). Heteroporosity of the capillary wall as indicated by cinematographic analysis of the passage of dyes. *Ann. N.Y. Acad. Sci.*, **116**, 765–773.

Landis, E. M. and Pappenheimer, J. R. (1963). Exchange of substances through the capillary walls. In *Handbook of Physiology*, Sect. 2, Circulation, Vol. 2, ed. W. F. Hamilton and P. Dow. Am. Physiol. Soc., Washington. pp. 961–1034.

Lane, B. P. and Rhodin, J. A. G. (1964). Cellular interrelationships and electrical activity in two types of smooth muscle. *J. Ultrastruct. Res.*, **10**, 470–488.

Lansing, A. I. (1954). Experimental studies on arteriosclerosis. In *Symposium on Atherosclerosis*. Publ. No. 338. Nat. Acad. Sci., Nat. Res. Council. pp. 50–60.

Lansing, A. I. (1959). Elastic tissue. In *The Arterial Wall*, ed. A. I. Lansing. Williams and Wilkins, Baltimore. pp. 136–160.

Lansing, A. I. (1961). Atherosclerosis and the nature of the arterial wall (Editorial). *Circulation*, **24**, 1283–1285.

Lansing, A. I., Rosenthal, T. B., Alex, M. and Dempsey, E. W. (1952). The structure and chemical characterization of elastic fibers as revealed by elastase and by electron microscopy. *Anat. Rec.*, **114**, 555–576.

Laschi, R. and Casanova, S. (1969). Fenestrae closed by a diaphragm in the endothelium of liver sinusoids. *J. Microscopie*, **8**, 1037–1040.

La Taillade, J. N., Gutstein, W. H. and Lazzarini-Robertson, A., Jr (1964). Study of experimental vasodilatation of rabbit abdominal aorta and its relationship to arterio-atherosclerosis. *J. Atheroscler. Res.*, **4**, 81–95.

Latta, H. (1970). The glomerular capillary wall. *J. Ultrastruct. Res.*, **32**, 526–544.

Latta, H., Maunsbach, A. B. and Cook, M. L. (1962). The juxtaglomerular apparatus as studied electron microscopically. *J. Ultrastruct. Res.*, **6**, 547–561.

Lautsch, E. V., McMillan, G. C. and Duff, G. L. (1953). Technics for the study of the normal and atherosclerotic arterial intima from its endothelial surface. *Lab. Invest.*, **2**, 397–407.

Leak, L. V. (1971). Frozen-fractured images of blood capillaries in heart tissue. *J. Ultrastruct. Res.*, **35**, 127–146.

Learoyd, B. M. and Taylor, M. G. (1966). Alterations with age in the viscoelastic properties of human arterial walls. *Circulation Res.*, **18**, 278–292.

Le Compte, P. M. (1967). Reactions of the vasa vasorum in vascular disease. In *Cowdry's Arteriosclerosis, a Survey of the Problem*, 2nd editn, ed. H. T. Blumenthal. C. Thomas, Springfield, Illinois. pp. 212–224.

Lee, K. T., Jones, R., Kim, D. N., Florentin, R., Coulston, F. and Thomas, W. A. (1966). Studies related to protein synthesis in experimental animals fed atherogenic diets. IV. ^{14}C-glycine incorporation into proteins of aortas and electron microscopy of non-necrotic atherosclerotic lesions of monkeys fed atherogenic diets for 8 to 16 months. *Exp. and Mol. Path.*, Suppl. 3, 108–123.

Lee, K. T., Lee, K. J., Lee, S. K., Imai, H. and O'Neal, R. M. (1970). Poorly differentiated subendothelial cells in swine aortas. *Exp. and Mol. Path.*, **13**, 118–129.

Leeson, T. S. (1971). Freeze-etch studies of rabbit eye. I. Choriocapillaris, lamina elastica (vitrea) and pigment epithelium. *J. Anat.*, **108**, 135–146.

Lesko, A., Babala, J. and Lojda, Z. (1966). Early histological and histochemical changes produced in the arterial wall of dogs by injecting blood clot. *Path. Microbiol.*, **29**, 279–284.

Levene, C. I. (1956a). The early lesions of atheroma in the coronary arteries. *J. Path. Bact.*, **72**, 79–82.

Levene, C. I. (1956b). The pathogenesis of atheroma of the coronary arteries. *J. Path. Bact.*, **72**, 83–86.

Levene, C. I. (1961). Collagen as a tensile component in the developing chick aorta. *Brit. J. exp. Path.*, **42**, 89–94.

Levene, C. I. and Poole, J. C. F. (1962). The collagen content of the normal and atherosclerotic human aortic intima. *Brit. J. exp. Path.*, **43**, 469–471.

Lever, J. D., Graham, J. D. P., Irvine, G. and Chick, W. J. (1965). The vesiculated axons in relation to arteriolar smooth muscle in the pancreas. A fine-structural and quantitative study. *J. Anat.*, **99**, 299–313.

Lever, J. D., Spriggs, T. L. B. and Graham, J. D. P. (1968). A formol-fluorescence, fine-structural and autoradiographic study of the adrenergic innervation of the vascular tree in the intact and sympathectomized pancreas of the cat. *J. Anat.*, **103**, 15–34.

Levick, J. R. and Michel, C. C. (1973). The permeability of individually perfused frog mesenteric capillaries to T1824 and T1824–albumin as evidence for a large pore system. *Quart. J. exp. Physiol.*, **58**, 67–85.

Lewis, G. P. (1964). Plasma kinins and other vasoactive compounds in acute inflammation. *Ann. N.Y. Acad. Sci.*, **116**, 847–854.

Lewis, G. P. and Matthews, J. (1970). The mechanism of functional vasodilatation in rabbit epigastric adipose tissue. *J. Physiol.*, **207**, 15–30.

Lewis, T. (1927). *The Blood Vessels of the Human Skin and their Responses*. Shaw and Sons, London.

Lewis, T. (1929). Observations upon the reactions of the vessels of the human skin to cold. *Heart*, **15**, 177–208.

Lewis, W. H. (1925–6). *The Transformation of Mononuclear Blood Cells into Macrophages, Epithelioid Cells and Giant Cells*. Harvey Lect., N.Y., Series 21, 77–110.

Lewis, W. H. (1931). The outgrowth of endothelium and capillaries in tissue culture. *Bull. Johns Hopk. Hosp.*, **48**, 242.

Lie, M., Sejersted, O. M. and Kiil, F. (1970). Local regulation of vascular cross section during changes in femoral arterial blood flow in dogs. *Circulation Res.*, **27**, 727–737.

Light, L. H. (1970). A recording spectrograph for analysing Doppler blood velocity signals (particularly from aortic flow) in real time. *J. Physiol.*, **207**, 42P–44P.

Lindquist, R. R., Guttmann, R. D. and Merril, J. P. (1968). Renal transplantation in the inbred rat. II. An immunohistochemical study of acute allograft rejection. *Am. J. Path.*, **52**, 531–545.

Ljung, B. (1970). Nervous and myogenic mechanisms in the control of a vascular neuroeffector system. *Acta physiol. scand.*, Suppl. 349, 33–68.

Loeven, W. A. (1969). Elastolytic enzymes in the vessel wall. *J. Atheroscler. Res.*, **9**, 35–45.

Longley, J. B., Banfield, W. G. and Brindley, D. C. (1960). Structure of the rete mirabile in the kidney of the rat as seen with the electron microscope. *J. biophys. biochem. Cytol.*, **7**, 103–105.

Lowy, J. and Hanson, J. (1962). Ultrastructure of invertebrate smooth muscles. *Physiol. Rev.*, Suppl. 5, 34–42.

Lowy, J. and Small, J. V. (1970). The organization of myosin and actin in vertebrate smooth muscle. *Nature*, **227**, 46–51.

Ludatscher, R. M. (1968). Fine structure of the muscular wall of rat pulmonary veins. *J. Anat.*, **103**, 345–357.

Ludatscher, R. M. and Stehbens, W. E. (1968). Fine structure of capillary angioma. *Am. J. Path.*, **52**, 25 (Abstract).

Luft, J. H. (1961). Improvements in Epoxy Resin embedding methods. *J. biophys. biochem. Cytol.*, **9**, 409–414.

Luft, J. H. (1965). The ultrastructural basis of capillary permeability. In *The Inflammatory Process*, ed. B. W. Zweifach, L. Grant and R. T. McCluskey. Academic Press, N.Y. and London. pp. 121–159.

Lund, B. and Myhre Jensen, O. (1970a). Renal transplantation in rabbits. II. Morphological alterations in autografts. *Acta path. microbiol. scand.*, A, **78**, 701–712.

Lund, B. and Myhre Jensen, O. (1970b). Renal transplantation in rabbits. III. Morphological alterations in allografts. *Acta path. microbiol. scand.*, A, **78**, 713–728.

Lutz, B. R. and Fulton, G. P. (1958). Smooth muscle and blood flow in small blood vessels. In *Factors Regulating Blood Flow*, ed. G. P. Fulton and B. Zweifach. Waverley Press, Baltimore. pp. 13–19.

McCombs, H. L., Zook, B. C. and McGandy, R. B. (1969). Fine structure of spontaneous atherosclerosis of the aorta in the squirrel monkey. *Am. J. Path.*, **55**, 235–252.

McConnell, J. G. and Roddie, I. C. (1970). A comparison of the behaviour of longitudinal and circular smooth muscle in bovine mesenteric vein. *J. Physiol.*, **207**, 82–83P.

McCuskey, R. S. (1966). A dynamic and static study of hepatic arterioles and hepatic sphincters. *Am. J. Anat.*, **119**, 455–478.

McCuskey, R. S. (1968). Dynamic microscopic anatomy of the fetal liver. III. Erythropoiesis. *Anat. Rec.*, **161**, 267–280.

McDonald, D. A. (1952a). The occurrence of turbulent flow in the rabbit aorta. *J. Physiol.*, **118**, 340–347.

McDonald, D. A. (1952b). The velocity of blood flow in the rabbit aorta studied with high-speed cinematography. *J. Physiol.*, **118**, 328–339.

McKay, D. and Hardaway, R. M. (1963). Thrombosis of arterioles, capillaries and venules: experimental considerations. In *The Peripheral Blood Vessels*, ed. J. L. Orbison and D. E. Smith. Williams and Wilkins, Baltimore. pp. 259–296.

McKay, D. G. and Margaretten, W. (1967). Disseminated intravascular coagulation in virus diseases. *Archs Int. Med.*, **120**, 129–152.

McPhaul, J. J., Jr and Dixon, F. J. (1969). Immunoreactive basement membrane antigens in normal human urine and serum. *J. exp. Med.*, **130**, 1395–1409.

Majno, G. (1965). Ultrastructure of the vascular membrane. In *Handbook of Physiology*, Sect. 2, Circulation, Vol. 3, ed. W. F. Hamilton and P. Dow. Am. Physiol. Soc., Washington. pp. 2293–2375.

Majno, G. (1970). Two endothelial 'novelties': endothelial contraction; collagenase digestion of the basement membrane. In *Vascular Factors and Thrombosis*, ed. K. M. Brinkhous *et al.* Schattauer Verlag, Stuttgart and N.Y. pp. 21–30.

Majno, G. and Palade, G. E. (1961). Studies on inflammation. I. The effect of histamine and serotonin on vascular permeability: an electron microscopic study. *J. biophys. biochem. Cytol.*, **11**, 571–605.

Majno, G., Shea, S. M. and Leventhal, M. J. (1969). Endothelial contraction induced by histamine-type mediators. *J. Cell Biol.*, **42**, 647–672.

Mancini, R. E., Barquet, J., Paz, M. A. and Vilar, O. (1965). Histoimmunological study of collagen during tendon fibrillogenesis. *Proc. Soc. exp. Biol. Med.*, **119**, 656–660.

Manion, W. C. (1963). Infectious angiitis. In *The Peripheral Blood Vessels*, ed. J. L. Orbison and D. E. Smith. Williams and Wilkins, Baltimore. pp. 221–231.

Manly, R. S. (ed.) (1970). *Adhesion in Biological Systems*. Academic Press, N.Y. and London.

Marbach, E. P. (1957). Collagen in wound healing. In *The Healing of wounds*, Symposium, ed. M. B. Williamson. McGraw–Hill Book Co. Inc., N.Y., Toronto, London.

Marchesi, V. T. (1961). The site of leucocyte emigration during inflammation. *Q. J. exp. Physiol.*, **46**, 115–118.

Marchesi, V. T. (1962). The passage of colloidal carbon through inflamed endothelium. *Proc. Roy. Soc.*, B, **156**, 550–552.

Marchesi, V. T. (1964). Some electron microscopic observations on interactions between leukocytes, platelets, and endothelial cells in acute inflammation. *Ann. N.Y. Acad. Sci.*, **116**, 774–788.

Marchesi, V. T. and Barrnett, R. J. (1963). The demonstration of enzymatic activity in pinocytic vesicles of blood capillaries with the electron microscope. *J. Cell Biol.*, **17**, 547–556.

Marchesi, V. T. and Florey, H. W. (1960). Electron micrographic observations on the emigration of leucocytes. *Q. J. exp. Physiol.*, **45**, 343–348.

Margaretten, W. and McKay, D. G. (1971). The requirement for platelets in the active Arthus reaction. *Am. J. Path.*, **64**, 257–270.

Mark, J. S. T. (1956). An electron microscope study of uterine smooth muscle. *Anat. Rec.*, **125**, 473–493.

Marshall, J. R. and O'Neal, R. M. (1966). The lipophage in hyperlipemic rats: an electron microscopic study. *Exp. and Mol. Path.*, **5**, 1–11.

Martin, G. R., Schiffman, E., Bladen, H. A. and Nylen, M. (1963). Chemical and morphological studies on the *in vitro* calcification of aorta. *J. Cell Biol.*, **16**, 243–252.

Maunsbach, A. (1966). The influence of different fixatives and fixation methods on the ultrastructure of rat kidney proximal tubule cells. II. Effects of varying the osmolality, ionic strength, buffer system and fixative concentration of glutaraldehyde solutions. *J. Ultrastruct. Res.*, **15**, 283–309.

Maximoff, A. (1917). Sur la culture *in vitro* du tissue lymphoïde des mammifères. *Compt. Rend. Soc. Biol.*, **80**, 222–225.

Maximov, A. (1916). The cultivation of connective tissue of adult mammals *in vitro*. *Arch. Russ. Anat., Hist. and Embryol.*, **i**, 105.

Maximow, A. (1928). Development of argyrophile and collagenous fibers in tissue culture. *Proc. Soc. exp. Biol. Med.*, **25**, 439–442.

Mayerson, H. S., Wolfram, C. G., Shirley, H. H. and Wasserman, K. (1960). Regional differences in capillary permeability. *Am. J. Physiol.*, **198**, 155–160.

Meadors, M. P., Jr and Johnson, W. W. (1970). Diffuse angio-endotheliosis. A fatal case in infancy with clinical and autopsy studies. *Archs Path.*, **90**, 572–576.

Mellander, S. and Johansson, B. (1968). Control of resistance, exchange and capacitance functions in the peripheral circulation. *Pharm. Revs.*, **20**, 117–196.

References

Mellander, S. and Lundvall, J. (1971). Role of tissue hyperosmalality in exercise hyperemia. *Circulation Res.*, **28**, Suppl. 1, 39–45.

Menkin, V. (1955). Factors concerned in the mobilization of leukocytes in inflammation. *Ann. N.Y. Acad. Sci.*, **59**, 956–985.

Menkin, V. (1960). Biochemical mechanisms in inflammation. *Brit. med. J.*, **1**, 1521–1528.

Menzies, D. W. and Roberts, J. T. (1963). Effect of age on the acidophilia of aortic elastin. *Nature*, **198**, 1006–1007.

Merkow, L., Lalich, J. L. and Angevine, D. M. (1961). Fibrillogenesis during repair of aortic injury. *Archs Path.*, **71**, 654–660.

Merrillees, N. C. R., Burnstock, G. and Holman, M. E. (1963). Correlation of fine structure and physiology of the innervation of smooth muscle in the guinea pig vas deferens. *J. Cell Biol.*, **19**, 529–550.

Michels, N. A. (1962). Hepatoportal–lienal circulation B. Gross anatomy. In *Blood Vessels and Lymphatics*, ed. D. I. Abramson. Academic Press, N.Y. and London. pp. 349–360.

Mikami S.-I., Oksche, A., Farner, D. S. and Vitums, A. (1970). Fine structure of the vessels of the hypophysial portal system of the Whitecrowned Sparrow, *Zonotrichia leucophrys gambelii*. *Z. Zellforsch. mikrosk. Anat.*, **106**, 155–174.

Mikata, A. and Niki, R. (1971). Permeability of postcapillary venules of the lymph node. An electron microscopic study. *Exp. and Mol. Path.*, **14**, 289–305.

Milhorat, T. H., Mosher, M. B., Hammock, M. K. and Murphy, C. F. (1970). Evidence for choroid–plexus absorption in hydrocephalus. *New Engl. J. Med.*, **283**, 286–289.

Mims, C. A. (1968). The response of mice to the intravenous injection of cowpox virus. *Brit. J. exp. Path.*, **49**, 24–32.

Moffat, D. B. (1967). The fine structure of the blood vessels of the renal medulla with particular reference to the control of the medullary circulation. *J. Ultrastruct. Res.*, **19**, 532–545.

Monie, I. W. (1945). Some observations on the subendothelial cushions of the umbilical arteries. *J. Anat.*, **79**, 137–144.

Moon, V. H. and Tershakovec, G. A. (1953). Dynamics of inflammation and of repair. III. Effects of tissue extracts and of protein split products upon capillary permeability and upon leucocytes. *Archs Path.*, **55**, 384–392.

Moore, D. H. and Ruska, H. (1957). The fine structure of capillaries and small arteries. *J. biophysic. biochem. Cytol.*, **3**, 457–462.

Moore, R. D. and Schoenberg, M. D. (1959). The relations of mucopolysaccharides of vessel walls to elastic fibres and endothelial cells. *J. Path. Bact.*, **77**, 163–169.

Moore, R. D., Schoenberg, M. D. and Koletsky, S. (1963). Cardiac lesions in experimental hypertension. *Archs Path.*, **75**, 28–44.

Morato, M. J. X. and Ferreira, J. F. D. (1957). Sur l'ultrastructure des capillaires de l'area postrema chez le lapin. *Compt. Rend. Soc. Biol.*, **151**, 1488–1490.

Moreno, A. H., Katz, A. I., Gold, L. D. and Reddy, R. V. (1970). Mechanics of distension of dog veins and other very thin-walled tubular structures. *Circulation Res.*, **27**, 1069–1080.

Moritz, A. R. (1941). The pathogenesis of arterial atrophy. *Am. J. Path.*, **17**, 597–598.

Moritz, A. R. (1967). Arteriosclerosis of the abdominal vessels. In '*Cowdry's Arteriosclerosis, a Survey of the Problem*', 2nd editn, ed. H. T. Blumental. C. Thomas, Springfield, Illinois. pp. 363–377.

Moses, J. M., Ebert, R. H., Graham, R. C. and Brine, K. L. (1964). Pathogenesis of inflammation. I. The production of an inflammatory substance from rabbit granulocytes *in vitro* and its relationship to leucocyte pyrogen. *J. exp. Med.*, **120**, 57–82.

Moss, N. S. and Benditt, E. P. (1970). Spontaneous and experimentally induced arterial lesions. I. An ultrastructural survey of the normal chicken aorta. *Lab. Invest.*, **22**, 166–183.
Movat, H. Z. and Fernando, N. V. P. (1962). The fine structure of connective tissue. I. The fibroblast. *Exp. and Mol. Path.*, **1**, 509–534.
Movat, H. Z. and Fernando, N. V. P. (1963). Allergic inflammation. I. The earliest fine structural changes at the blood–tissue barrier during antigen–antibody interaction. *Am. J. Path.*, **42**, 41–59.
Muir, A. R. and Peters, A. (1962). Quintuple-layered membrane junctions at terminal bars between endothelial cells. *J. Cell Biol.*, **12**, 443–448.
Munger, B. L. (1961). The ultrastructure and histophysiology of human eccrine sweat glands. *J. biophys. biochem. Cytol.*, **11**, 385–402.
Murray, M., Schrodt, G. R. and Berg, H. F. (1966). Role of smooth muscle cells in healing of injured arteries. *Archs Path.*, **82**, 138–146.
Neuman, R. E. and Logan, M. A. (1950). The determination of collagen and elastin in tissues. *J. Biol. Chem.*, **186**, 549–556.
Nichol, J., Girling, F., Jerrard, W., Claxton, E. B. and Burton, A. C. (1951). Fundamental instability of the small blood vessels and critical closing pressures in vascular beds. *Am. J. Physiol.*, **164**, 330–344.
Nicoll, P. A. and Frayser, R. (1967). Physiological considerations of the microcirculation as related to shock. *Progress in Cardiovascular Diseases*, **9**, 558–570.
Nicoll, P. A. and Webb, R. L. (1946). Blood circulation in the subcutaneous tissue of the living bat's wing. *Ann. N.Y. Acad. Sci.*, **46**, 697–711.
Nicoll, P. A. and Webb, R. L. (1955). Vascular patterns and active vasomotion as determiners of flow through minute vessels. *Angiology*, **6**, 291–308.
Nishimaru, Y. (1969). Body fluid flow in tissue spaces. *Hiroshima J. Med. Sci.*, **18**, 185–196.
North, R. J. (1966a). The localisation by electron microscopy of nucleoside phosphatase activity in guinea pig phagocytic cells. *J. Ultrastruct. Res.*, **16**, 83–95.
North, R. J. (1966b). The localisation by electron microscopy of acid phosphatase activity in guinea pig macrophages. *J. Ultrastruct. Res.*, **16**, 96–108.
North, R. J. (1967). Structure and function in phagocytic cells. Thesis submitted to the Australian National University for the degree of Doctor of Philosophy.
North, R. J. and Mackaness, G. B. (1963). Electronmicroscopical observation on the peritoneal macrophages of normal mice and mice immunised with *Listeria monocytogenes*. I. Structure of normal macrophages and the early cytoplasmic response to the presence of ingested bacteria. *Brit. J. exp. Path.*, **44**, 601–607.
Northover, A. M. and Northover, B. J. (1970). The effect of vasoactive substances on rat mesenteric blood vessels. *J. Path.*, **101**, 99–108.
Nour-Eldin, F. (1966). A study on the evolvement of blood-clotting activity in the human body and its role in thrombosis and factors VIII and IX defects. *Ann. N.Y. Acad. Sci.*, **136**, 219–258.
Novikoff, A. B. (1961). Lysosomes and related particles. In *The Cell, Biochemistry, Physiology, Morphology*, Vol. 2, ed. J. Brachet and A. E. Mirsky. Academic Press, N.Y. and London. pp. 423–488.
Novikoff, A. B. (1963). Lysosomes in the physiology and pathology of cells. Contributions of staining methods. In *Ciba Foundation Symposium on Lysosomes*, ed. A. V. S. de Reuck and M. P. Cameron. J. and A. Churchill Ltd, London. pp. 36–77.
Nyström, S. H. M. (1963). Development of intracranial aneurysms as revealed by electron microscopy. *J. Neurosurg.*, **20**, 329–337.
Oberling, C. (1959). The structure of cytoplasm. *Int. Rev. Cytol.*, **8**, 1–31.
Ohta, G., Sasaki, H., Matsubara, F., Tanishima, K. and Watanabe, S. (1962).

Heparin-like substances in cement lines of vascular endothelium of guinea pigs. *Proc. Soc. exp. Biol. Med.*, **109**, 298–300.

Olsen, B. R. (1963). Electron microscope studies on collagen. II. Mechanism of linear polymerization of tropocollagen molecules. *Z. Zellforsch.*, **59**, 199–213.

O'Neal, R. M., Jordan, G. L., Jr, Rabin, E. R., De Bakey, M. E. and Halpert, B. (1964). Cells grown on isolated intravascular dacron hub, an electron microscopic study. *Exp. and Mol. Path.*, **3**, 403–412.

Oosaki, T. and Ishii, S. (1964). Junctional structure of smooth muscle cells. The ultrastructure of the regions of junction between smooth muscle cells in the rat small intestine. *J. Ultrastruct. Res.*, **10**, 567–577.

Opdyke, D. F. (1970). Hemodynamics of blood flow through the spleen. *Am. J. Physiol.*, **219**, 102–106.

Packham, M. A., Rowsell, H. C., Jørgensen, L. and Mustard, J. F. (1967). Localized protein accumulation in the wall of the aorta. *Exp. and Mol. Path.*, **7**, 214–232.

Palade, G. E. (1952). A study of fixation for electron microscopy. *J. exp. Med.*, **95**, 285–298.

Palade, G. E. (1953a). Fine structure of blood capillaries. *J. appl. Physics*, **24**, 1424.

Palade, G. E. (1953b). An electron microscope study of the mitochondrial structure. *J. Histochem. Cytochem.*, **1**, 188–211.

Palade, G. E. (1955). Relations between the endoplasmic reticulum and the plasma membrane in macrophages. *Anat. Rec.*, **121**, 445.

Palade, G. E. (1960). Transport in quanta across the endothelium of blood capillaries. *Anat. Rec.*, **136**, 254 (Abstract).

Palade, G. E. and Porter, K. R. (1954). Studies on the endoplasmic reticulum. I. Its identification in cells *in situ*. *J. exp. Med.*, **100**, 641–656.

Palay, S. L. and Karlin, L. J. (1959). An electron microscopic study of the intestinal villus. I. The fasting animal. *J. biophys. biochem. Cytol.*, **5**, 363–372.

Panner, B. J. and Honig, C. R. (1970). Locus and state of aggregation of myosin in tissue sections of vertebrate smooth muscle. *J. Cell Biol.*, **44**, 52–61.

Pappas, G. D. and Tennyson, V. M. (1962). An electron microscopic study of the passage of colloidal particles from the blood vessels of the ciliary processes and choroid plexus of the rabbit. *J. Cell Biol.*, **15**, 227–239.

Pappenheimer, J. R., Renkin, E. M. and Borrero, L. M. (1951). Filtration, diffusion and molecular sieving through peripheral capillary membranes. A contribution to the pore theory of capillary permeability. *Am. J. Physiol.*, **167**, 13–46.

Parker, F. (1958). An electron microscope study of coronary arteries. *Am. J. Anat.*, **103**, 247–275.

Parker, F. and Odland, G. F. (1966a). A correlative histochemical, biochemical and electron microscopic study of experimental atherosclerosis in the rabbit aorta with special reference to the myo-intimal cell. *Am. J. Path.*, **48**, 197–239.

Parker, F. and Odland, G. F. (1966b). A light microscopic, histochemical and electron microscopic study of experimental atherosclerosis in rabbit coronary artery and a comparison with rabbit aorta atherosclerosis. *Am. J. Path.*, **48**, 451–481.

Parpart, A. K., Whipple, A. O. and Chang, J. J. (1955). The micro-circulation of the spleen of the mouse. *Angiology*, **6**, 350–368.

Partridge, S. M. (1962). Elastin. *Adv. Protein Chem.*, **17**, 227–302.

Partridge, S. M. (1966). Elastin. In *The Physiology and Biochemistry of Muscle as a Food*, ed. E. J. Briskey, R. G. Cassens and J. C. Trautman. University of Wisconsin Press, Madison, Milwaukee and London. pp. 327–339.

Partridge, S. M. and Davis, H. F. (1955). The chemistry of connective tissues. III. Composition of the soluble proteins derived from elastin. *Biochem. J.*, **61**, 21–30.

Partridge, S. M., Elsden, D. F. and Thomas, J., and Dorfman, A., Telser, A. and

Ho, P.-L. (1964). Biosynthesis of the desmosine and isodesmosine cross-bridges in elastin. *Biochem. J.*, **93**, 30c–33c.
Paterson, J. C., Slinger, S. J. and Gartley, K. M. (1948). Experimental coronary sclerosis. *Archs Path.*, **45**, 306–318.
Paule, W. J. (1963). Electron microscopy of the newborn rat aorta. *J. Ultrastruct. Res.*, **8**, 219–235.
Peachey, L. D. (1964). Electron microscopic observations on the accumulation of divalent cations in intramitochondrial granules. *J. Cell Biol.*, **20**, 95–111.
Pearse, A. G. E. (1960). *Histochemistry, Theoretical and Applied*, 2nd editn. J. and A. Churchill Ltd, London.
Pease, D. C. and Molinari, S. (1960). Electron microscopy of muscular arteries: Pial vessels of the cat and monkey. *J. Ultrastruct. Res.*, **3**, 447–468.
Pease, D. C. and Paule, W. J. (1960). Electron microscopy of elastic arteries; the thoracic aorta of the rat. *J. Ultrastruct. Res.*, **3**, 469–483.
Pease, D. C., Molinari, S. and Kershaw, T. (1958). An electron microscopic study of the larger intracranial blood vessels of cats and dogs. *Anat. Rec.*, **130**, 355.
Pedersen, N. C. and Morris, B. (1970). The role of the lymphatic system in the rejection of homografts: a study of lymph from renal transplants. *J. exp. Med.*, **131**, 936–969.
Pentreath, V. W. and Cottrell, G. A. (1970). The blood supply to the central nervous system of *Helix pomatia*. *Z. Zellforsch. mikrosk. Anat.*, **111**, 160–178.
Peracchia, C. and Mittler, B. S. (1972). Fixation by means of glutaraldehyde–hydrogen peroxide reaction products. *J. Cell Biol.*, **53**, 234–238.
Peterson, L. H. (1962). Properties and behaviour of living vascular wall. *Physiol. Rev.*, **42**, Suppl. 5, 309–324.
Peterson, L. H., Jensen, R. E. and Parnell, J. (1960). Mechanical properties of arteries *in vivo*. *Circulation Res.*, **8**, 622–639.
Phelps, P. C. and Luft, J. H. (1969). Electron microscopical study of relaxation and constriction in frog arterioles. *Am. J. Anat.*, **125**, 399–428.
Picken, L. E. R. (1940). The fine structure of biological systems. *Biol. Rev.*, **15**, 133–167.
Piez, K. A., Weiss, E. and Lewis, M. S. (1960). The separation and characterization of the α- and β-components of calf skin collagen. *J. Biol. Chem.*, **235**, 1987–1991.
Policard, A. and Collet, A. (1958). Les données nouvelles apportées par la microscopie électronique à la connaissance du fonctionnement des capillaires sanguins. *Revue fr. Etud. clin. biol.*, **3**, 205–207.
Policard, A., Collet, A. and Giltaire-Ralyte, L. (1955). Observations au microscope électronique sur la structure infra microscopique des artérioles des mammifères. *Bull. Microsc. appl.*, **5**, 3–4.
Policard, A., Collet, A. and Pregermain, S. (1957). Etude au microscope electronique des capillaires pulmonaires. *Acta anat.*, **30**, 624–638.
Poole, J. C. F., Cromwell, S. B. and Benditt, E. P. (1970). Behaviour of smooth muscle cells and formation of extracellular structures in the reaction of arterial walls to injury. *Am. J. Path.*, **62**, 391–413.
Poole, J. C. F. and Florey, H. W. (1958). Changes in the endothelium of the aorta and the behaviour of macrophages in experimental atheroma of rabbits. *J. Path. Bact.*, **75**, 245–252.
Poole, J. C. F., Sanders, A. G. and Florey, H. W. (1958). The regeneration of aortic endothelium. *J. Path. Bact.*, **75**, 133–143.
Porter, K. R. (1953). Observations on a submicroscopic basophilic component of cytoplasm. *J. exp. Med.*, **97**, 727–749.
Porter, K. R. and Kallman, F. (1953). The properties and effects of osmium tetroxide

as a tissue fixative with special reference to its use for electron microscopy. *Expl Cell Res.*, **4**, 127–141.

Prosser, C. L. (1950). Circulation of body fluids. In *Comparative Animal Physiology*, ed. C. L. Prosser. W. B. Saunders Co., Philadelphia and London. pp. 531–575.

Quick, A. J. (1957). *Hemorrhagic Diseases*. Lea and Febiger, Philadelphia.

Ramsey, E. M. (1955). Vascular patterns in the endometrium and the placenta. *Angiology*, **6**, 321–339.

Ramsey, E. M. (1962). Placental circulation. In *Blood Vessels and Lymphatics*, ed. D. I. Abramson. Academic Press, N.Y. and London. pp. 465–485.

Randall, J. T., Fraser, R. D. B., Jackson, S., Martin, A. V. W. and North, A. C. T. (1952). Aspects of collagen structure. *Nature*, **169**, 1029–1033.

Raper, A. J., Kontos, H. A. and Patterson, J. L., Jr (1971). Response of pial precapillary vessels to changes in arterial carbon dioxide tension. *Circulation Res.*, **28**, 518–523.

Rees, P. M. and Jepson, P. (1970). Measurement of arterial geometry and wall composition in the carotid sinus baroreceptor area. *Circulation Res.*, **26**, 461–467.

Reuben, S. R., Swadling, J. P. and Lee, G. de J. (1970). Velocity profiles in the main pulmonary artery of dogs and man, measured with a thin-film resistance anemometer. *Circulation Res.*, **27**, 995–1001.

Rhodin, J. A. G. (1962*a*). Fine structure of vascular walls in mammals with special reference to smooth muscle component. *Physiol. Rev.*, **42**, Suppl. 5, 48–87.

Rhodin, J. A. G. (1962*b*). The diaphragm of capillary endothelial fenestrations. *J. Ultrastruct. Res.*, **6**, 171–185.

Rhodin, J. A. G. (1967). The ultrastructure of mammalian arterioles and precapillary sphincters. *J. Ultrastruct. Res.*, **18**, 181–223.

Rhodin, J. A. G. (1968). Ultrastructure of mammalian venous capillaries, venules, and small collecting veins. *J. Ultrastruct. Res.*, **25**, 452–500.

Rhodin, J. and Dalhamn, T. (1955). Electron microscopy of collagen and elastin in lamina propria of the tracheal mucosa of rat. *Expl Cell Res.*, **9**, 371–375.

Rivkin, L. M., Friedman, M. and Byers, S. O. (1963). Thrombo-atherosclerosis in aortic venous autografts. A comparative study. *Brit. J. exp. Path.*, **44**, 16–23.

Roach, M. R. (1970). Role of vascular wall elastic tissue in hemostasis. In *Vascular Factors and Thrombosis*, ed. K. M. Brinkhous *et al.* Schattauer Verlag, Stuttgart and N.Y. pp. 59–77.

Robbins, E. (1961). Some theoretical aspects of osmium tetroxide fixation with special reference to the metaphase chromosomes of cell cultures. *J. biophys. biochem. Cytol.*, **11**, 449–455.

Robertson, H. F. (1929). Vascularization of the thoracic aorta. *Archs Path.*, **8**, 881–893.

Robertson, J. D. (1959). The ultrastructure of cell membranes and their derivatives. *Biochem. Soc. Symp.*, **16**, 3–43.

Rocha, H. and Fekety, F. R. (1964). Acute inflammation in the renal cortex and medulla following thermal injury. *J. exp. Med.*, **119**, 131–137.

Rodbard, S. (1971). The burden of the resistance vessels. *Circulation Res.*, **28**, Suppl. 1, 2–8.

Rolewicz, T. F., Gisslen, J. L. and Zimmerman, B. G. (1970). Uptake of norepinephrine-^{3}H by cutaneous arteries and veins of the dog. *Am. J. Physiol.*, **219**, 62–67.

Romer, A. S. (1950). *The Vertebrate Body*. W. B. Saunders Co., Philadelphia and London.

Rosenblum, W. I. (1970). Effects of blood pressure and blood viscosity on fluorescein transit time in the cerebral microcirculation in the mouse. *Circulation Res.*, **27**, 825–833.

Rosenbluth, J. and Wissig, S. L. (1964). The distribution of exogenous ferritin in toad spinal ganglia and the mechanism of its uptake by neurons. *J. Cell Biol.*, **23**, 307–325.

Ross, R. (1971). The smooth muscle cell. II. Growth of smooth muscle in culture and formation of elastic fibers. *J. Cell Biol.*, **50**, 172–186.

Ross, R. and Benditt, E. P. (1961). Wound healing and collagen formation. I. Sequential changes in components of guinea pig skin wounds observed in the electron microscope. *J. biophys. biochem. Cytol.*, **11**, 677–700.

Ross, R. and Benditt, E. P. (1962). Wound healing and collagen formation. III. A quantitative radioautographic study of the utilization of proline-H^3 in wounds from normal and scorbutic guinea pigs. *J. Cell Biol.*, **15**, 99–108.

Ross, R. and Bornstein, P. (1969). The elastic fibre. I. The separation and partial characterization of its macromolecular components. *J. Cell Biol.*, **40**, 366–381.

Ross, R. and Klebanoff, S. J. (1971). The smooth muscle cell. I. *In vivo* synthesis of connective tissue proteins. *J. Cell Biol.*, **50**, 159–171.

Ross, R. and Lillywhite, J. W. (1965). The fate of buffy coat cells grown in subcutaneously implanted diffusion chambers. *Lab. Invest.*, **14**, 1568–1585.

Rothbard, S. and Watson, R. F. (1961). Antigenicity of rat collagen. Demonstration of antibody to rat collagen in the renal glomeruli of rats by fluorescence microscopy. *J. exp. Med.*, **113**, 1041–1052.

Rowley, D. A. (1964). Venous constriction as the cause of increased vascular permeability produced by 5-hydroxytryptamine, histamine, bradykinin and 48/80 in the rat. *Brit. J. exp. Path.*, **45**, 56–67.

Rüegg, J. C. (1971). Smooth muscle tone. *Physiol. Rev.*, **51**, 201–248.

Ryan, G. B., Cliff, W. J., Gabbiani, G., Irlé, C., Statkov, P. R. and Majno, G. (1973). Myofibroblasts in an avascular fibrous tissue. *Lab. Invest.* **29**, 197–206.

Sabatini, D. D., Bensch, K. and Barrnett, R. J. (1963). Cytochemistry and electron microscopy. The preservation of cellular ultrastructure and enzymatic activity by aldehyde fixation. *J. Cell Biol.*, **17**, 19–58.

Sabin, F. R. (1921). Studies on blood; the vitally stainable granules as a specific criterion for erythroblasts and the differentiation of the three strains of the white blood cells as seen in the living chick's yolk sac. *Johns Hopk. Hosp. Bull.*, **32**, 314.

Sainte-Marie, G. (1966). The postcapillary venules in the mediastinal lymph node of ten-week-old rats. *Revue Can. Biol.*, **25**, 263–284.

Salgado, E. D. (1970). Medial aortic lesions in rats with metacorticoid hypertension. *Am. J. Path.*, **58**, 305–327.

Samuelson, A., Becker, A. E. and Wagenvoort, C. A. (1970). A morphometric study of pulmonary veins in normal infants and infants with congenital heart disease. *Archs Path.*, **90**, 112–116.

Sanders, A. G., Ebert, R. H. and Florey, H. W. (1940). The mechanism of capillary contraction. *Q. J. exp. Physiol.*, **30**, 281–287.

Sandison, J. C. (1924). A new method for the microscopic study of living growing tissues by the introduction of a transparent chamber in the rabbit's ear. *Anat. Rec.*, **28**, 281–287.

Sandison, J. C. (1932). Contraction of blood vessels and observations on the circulation in the transparent chamber in the rabbit's ear. *Anat. Rec.*, **54**, 105.

Schallock, G. (1938). Über die Entwicklung und das weitere Schicksal der Nabelarterie. Zugleich ein Beitrag zur Gefässpathologie. *Virchows Arch. path. Anat. Physiol.*, **302**, 195–209.

Schaper, W. (1971). DNA synthesis and mitoses in coronary collateral vessels of the dog. *Circulation Res.*, **28**, 671–679.

Scharfstein, H., Gutstein, W. H. and Lewis, L. (1963). Changes of boundary layer flow in model systems: implications for initiation of endothelial injury. *Circulation Res.*, **13**, 580–584.

Schlichter, J. and Harris, R. (1949). The vascularization of the aorta. II. A comparative study of the aortic vascularization of several species in health and disease. *Am. J. med. Sci.*, **218**, 610–615.

Schlichter, J. G., Katz, L. N. and Meyer, J. (1949). The occurrence of atheromatous lesions after cauterization of the aorta followed by cholesterol administration. *Am. J. med. Sci.*, **218** 603–609

Schloss, G. and Schumacker, H. B., Jr (1950). Studies in vascular repair. IV. The use of free vascular transplants for bridging arterial defects. An historical review with particular reference to histological observations. *Yale J. Biol. Med.*, **22**, 273–290.

Schmidt-Diedrichs, A. and Courtice, F. C. (1963). The removal of various lipoproteins from doubly-ligated segments of artery and vein in the rabbit. *Brit. J. exp. Path.*, **44**, 345–350.

Schmitt, F. O. and Gross, J. (1948). Further progress in the electron microscopy of collagen. *J. Am. Leather Chem. Ass.*, **43**, 658–675.

Schneider, W. C. (1959). Mitochondrial metabolism. *Adv. Enzymol.*, **21**, 1–72.

Schoefl, G. I. (1963). Studies on inflammation. III. Growing capillaries: their structure and permeability. *Virchows Arch. path. Anat. Physiol.*, **337**, 97–141.

Schoefl, G. I. (1970). Structure and permeability of venules in lymphoid tissue. *7th International Congress of Electron Microscopy*, Grenoble (1970), Vol. 3, 589–590.

Schoefl, G. I. (1972). The migration of lymphocytes across the vascular endothelium in lymphoid tissue. *J. exp. Med.*, **136**, 568–584.

Schoefl, G. I. and French, J. E. (1968). Vascular permeability to particulate fat: morphological observations on vessels of lactating mammary gland and of lung. *Proc. Roy. Soc.*, B, **169**, 153–165.

Scholander, P. F. and Schevill, W. E. (1955). Counter-current vascular heat exchange in the fins of whales. *J. appl. Physiol.*, **8**, 279–282.

Scholander, P. F., Walters, V., Hock, R. and Irving, L. (1950). Body insulation of some arctic and tropical mammals and birds. *Biol. Bull.*, **99**, 225–236.

Schwarz, W. and Dettmer, N. (1953). Elektronenmikroskopische Untersuchung des elastichen Gewebes in der Media der menschlichen Aorta. *Virchows Arch. path. Anat. Physiol.*, **323**, 243–268.

Scott, J. B. and Radawski, D. (1971). Role of hyperosmolarity in the genesis of active and reactive hyperemia. *Circulation Res.*, **28**, Suppl. 1, 26–32.

Scott, R. F., Jones, R., Daoud, A. S., Zumbo, O., Coulston, F. and Thomas, W. A. (1967). Experimental atherosclerosis in Rhesus monkeys. II. Cellular elements of proliferative lesions and possible role of cytoplasmic degeneration in pathogenesis as studied by electron microscopy. *Exp. and Mol. Path.*, **7**, 34–57.

Seifert, K. (1963). Elektronenmikroskopische Untersuchungen der Aorta des Kaninchens. *Z. Zellforsch. mikrosk. Anat.*, **60**, 293–312.

Selkurt, E. E. (1962). Splenic circulation. In *Blood Vessels and Lymphatics*, ed. D. I. Abramson. Academic Press, N.Y. and London. pp. 369–371.

Sengel, A. and Stoebner, P. (1970). Golgi origin of tubular inclusions in endothelial cells. *J. Cell Biol.*, **44**, 223–226.

Serafini-Fracassini, A. and Tristram, G. R. (1966). Electron microscopic study and amino acid analysis on human aortic elastin. *Proc. Roy. Soc. Edinb.*, B, **69**, 334–344.

Sethi, N. and Brookes, M. (1971). Ultrastructure of the blood vessels in the chick allantois and chorioallantois. *J. Anat.*, **109**, 1–15.

References

Shea, S. M. and Karnovsky, M. J. (1966). Brownian motion: a theoretical explanation for the movement of vesicles across the endothelium. *Nature*, **212**, 353–355.
Shea, S. M., Karnovsky, M. J. and Bossert, W. H. (1969). Vesicular transport across endothelium: simulation of a diffusion model. *J. theor. Biol.*, **24**, 30–42.
Shenkin, H. A. and Bouzarth, W. F. (1970). Clinical methods of reducing intracranial pressure role of the cerebral circulation. *N. Engl. J. Med.*, **282**, 1465–1471.
Short, D. (1966). Morphology of the intestinal arterioles in chronic human hypertension. *Brit. Heart J.*, **28**, 184–192.
Siggins, G. R. and Bloom, F. E. (1970). Cytochemical and physiological effects of 6-hydroxydopamine on periarteriolar nerves of frogs. *Circulation Res.*, **27**, 23–38.
Siller, W. G. (1962). Two cases of aortic rupture in fowls. *J. Path. Bact.*, **83**, 527–533.
Silva, D. G. and Ikeda, M. (1971). Ultrastructural and acetylcholinesterase studies on the innervation of the ductus arteriosus, pulmonary trunk and aorta of the fetal lamb. *J. Ultrastruct. Res.*, **34**, 358–374.
Simons, J. R. and Michaelis, A. R. (1953). A cinematographic technique, using ultra-violet illumination, for amphibian blood circulation. *Nature*, **171**, 801.
Simpson, C. F. and Harms, R. H. (1964). Pathology of the aorta of chicks fed a copper-deficient diet. *Exp. and Mol. Path.*, **3**, 390–400.
Simpson, F. O. and Devine, C. E. (1966). The fine structure of autonomic neuromuscular contacts in arterioles of sheep renal cortex. *J. Anat.*, **100**, 127–137.
Smith, C., Seitner, M. M. and Wang, H.P. (1951). Aging changes in the tunica media of the aorta. *Anat. Rec.*, **109**, 13–39.
Smith, D. E. (1963). The tissue mast cell. *Int. Rev. Cytol.*, **14**, 327–386.
Smith, J. B., McIntosh, G. H. and Morris, B. (1970). The traffic of cells through tissues: a study of peripheral lymph in sheep. *J. Anat.*, **107**, 87–100.
Smith, J. W. (1967). Microsurgery and vasa vasorum. In *Microvascular Surgery*, ed. R. M. P. Donaghy and M. G. Yaşargil. Georg Thieme Verlag, Stuttgart; C. V. Mosby Co., St Louis. pp. 57–63.
Snodgrass, M. J. and Snook, T. (1971). A study of some histochemical and phagocytic reactions of the reticuloendothelial system of the rabbit spleen. *Anat. Rec.*, **170**, 243–253.
Sohal, R. S. and Burch, G. E. (1969). Electron microscopic study of the endocardium in Coxsackie virus B_4 infected mice. *Am. J. Path.*, **55**, 133–145.
Somlyo, A. V. and Somlyo, A. P. (1964). Vasomotor function of smooth muscle in the main pulmonary artery. *Am. J. Physiol.*, **206**, 1196–1200.
Somlyo, A. P. and Somlyo, A. V. (1968). Vascular smooth muscle. I. Normal structure, pathology, biochemistry and biophysics. *Pharmac. Rev.*, **20**, 197–272.
Somlyo, A. P. and Somlyo, A. V. (1970). Vascular smooth muscle. II. Pharmacology of normal and hypertensive vessels. *Pharmac. Rev.*, **22**, 249–353.
Somlyo, A. V., Woo, C.-Y. and Somlyo, A. P. (1965). Responses of nerve-free vessels to vasoactive amines and polypeptides. *Am. J. Physiol.*, **208**, 748–753.
Sorokin, S. (1962). Centrioles and the formation of rudimentary cilia by fibroblasts and smooth muscle cells. *J. Cell Biol.*, **15**, 363–377.
Spain, D. M. and Aristizabal, N. (1962). Rabbit local tissue response to triglycerides, cholesterol, and its ester. *Archs Path.*, **73**, 82–85.
Spector, W. G. and Willoughby, D. A. (1964). Vasoactive amines in acute inflammation. *Ann. N.Y. Acad. Sci.*, **116**, 838–846.
Spraragen, S. C., Bond, V. P. and Dahl, L. K. (1962). Role of hyperplasia in vascular lesions of cholesterol-fed rabbits studied with thymidine-H^3 autoradiography. *Circulation Res.*, **11**, 329–336.
Städeli, H. (1966). Der Einfluss des Lebensalters auf die Wandstruktur der isolierten, künstlich gedehnten Aorta thoracalis des Meerschweinchens. *Angiologica*, **3**, 213–225.

References

Staehelin, L. A. (1968). The interpretation of freeze-etched artificial and biological membranes. *J. Ultrastruct. Res.*, **22**, 326–347.

Starling, E. H. (1896). On the absorption of fluids from connective tissue spaces. *J. Physiol.*, **19**, 312–326.

Stearner, S. P. and Sanderson, M. H. (1971). Microvasculature and loose connective tissue in freeze-fracture preparations. *Z. Zellforsch. mikrosk. Anat.*, **114**, 301–308.

Stehbens, W. E. (1959). Turbulence of blood flow. *Q. J. exp. Physiol.*, **44**, 110–117.

Stehbens, W. E. (1962). The production of sulphated mucopolysaccharides by endothelium. *J. Path. Bact.*, **83**, 337–345.

Stehbens, W. E. (1965). Ultrastructure of vascular endothelium in the frog. *Q. J. exp. Physiol.*, **50**, 375–384.

Stehbens, W. E. and Ludatscher, R. M. (1968). Fine structure of senile angiomas of human skin. *Angiology*, **19**, 581–592.

Stehbens, W. E. and Meyer, E. (1965). Ultrastructure of endothelium of the frog heart. *J. Anat.*, **99**, 127–134.

Stein, O., Eisenberg, S. and Stein, Y. (1969). Aging of aortic smooth muscle cells in rats and rabbits. A morphologic and biochemical study. *Lab. Invest.*, **21**, 386–397.

Stemerman, M. B., Baumgartner, H. R. and Spaet, T. H. (1971). The subendothelial microfibril and platelet adhesion. *Lab. Invest.*, **24**, 179–186.

Stetten, M. R. (1949). Some aspects of the metabolism of hydroxyproline studied with the aid of isotopic nitrogen. *J. biol. Chem.*, **181**, 31–37.

Still, W. J. S. (1964). Pathogenesis of experimental atherosclerosis. *Archs Path.*, **78**, 601–612.

Still, W. J. S. and Dennison, S. M. (1967). Reaction of the arterial intima of the rabbit to trauma and hyperlipemia. *Exp. and Mol. Path.*, **6**, 245–253.

Still, W. J. S. and O'Neal, R. M. (1962). Electron microscopic study of experimental atherosclerosis in the rat. *Am. J. Path.*, **40**, 21–35.

Stoeckenius, W., Schulman, J. H. and Prince, L. M. (1960). The structure of myelin figures and microemulsions as observed with the electron microscope. *Kolloidzeitschrift*, **169**, 170–180.

Strauss, E. W. (1963). The absorption of fat by intestine of golden hamster *in vitro*. *J. Cell Biol.*, **17**, 597–607.

Striker, G. E. and Smuckler, E. A. (1970). An ultrastructural study of glomerular basement membrane synthesis. *Am. J. Path.*, **58**, 531–555.

Strum, J. M. and Karnovsky, M. J. (1970). Cytochemical localization of endogenous peroxidase in thyroid follicular cells. *J. Cell Biol.*, **44**, 655–666.

Stump, M. M., Jordan, G. L., Jr, De Bakey, M. E. and Halpert, B. (1963). Endothelium grown from circulating blood on isolated intravascular dacron hub. *Am. J. Path.*, **43**, 361–367.

Sutter, M. C. (1965). The pharmacology of isolated veins. *Brit. J. Pharmacol.*, **24**, 742–751.

Švejcar, J., Přerovský, I., Linhart, J. and Kruml, J. (1964). Biochemical differences in the composition of primary varicose veins. *Am. Heart J.*, **67**, 572–574.

Takada, M. (1970). Fenestrated venules of the large salivary glands. *Anat. Rec.*, **166**, 605–610.

Takada, M. and Gore, I. (1968). Capillary endothelial fenestrations in the lamina propria of the guinea pig tongue. *Anat. Rec.*, **161**, 465–470.

Tao, K.-Y. T., Treadwell, C. R., Hilker, D. M. and McGavack, T. H. (1962). Effect of age upon synthesis and turnover of connective tissue proteins in rats. In *Biological Aspects of Aging*, ed. N. W. Shock. Columbia University Press, N.Y. and London. pp. 343–347.

Tauchi, H. (1962). Mechanism of senile atrophy – histopathological, micrometrical, and electron microscopical studies. In *Biological Aspects of Aging*, ed. N. W. Shock. Columbia University Press, N.Y. and London. pp. 157–162.

Taylor, C. B. (1955). The reactions of arteries to injury by physical agents with a discussion of factors influencing arterial repair. In *Symposium on Atherosclerosis*. Publ. No. 338, Nat. Acad. Sci., Nat. Res. Council, pp. 74–90.

Taylor, C. B., Baldwin, D. and Hass, G. M. (1950). Localized arteriosclerotic lesions induced in the aorta of the juvenile rabbit by freezing. *Archs Path.*, **49**, 623–640.

Taylor, H. E., Shepherd, W. E. and Robertson, C. E. (1961). An immunohistochemical examination of granulation tissue with glomerular and lung antiserums. *Am. J. Path.*, **38**, 39–48.

Tennent, D. M., Zanetti, M. E., Ott, W. H., Kuron, G. W. and Siegel, H. (1956). Influence of crystalline elastase on experimental atherosclerosis in the chicken. *Science*, **124**, 588.

Terzakis, J. A. (1963). The ultrastructure of normal human first trimester placenta. *J. Ultrastruct. Res.*, **9**, 268–284.

Texon, M. (1957). A hemodynamic concept of atherosclerosis, with particular reference to coronary occlusion. *A.M.A. Archs Int. Med.*, **99**, 418–427.

Thaemert, J. C. (1963). The ultrastructure and disposition of vesiculated nerve processes in smooth muscle. *J. Cell Biol.*, **16**, 361–377.

Thoma, R. (1893). Üntersuchungen über die Histogenese und Histomechanik des Gefasssystems. Stuttgart.

Thoma, R. (1896). *Textbook of General Pathology and Pathological Anatomy*. Translated A. Bruce. Adam and Charles Black, London.

Thurau, K. (1971). Micropuncture evaluation of local control of arteriolar resistance of kidney and brain. *Circulation Res.*, **28**, Suppl. 1, 106–114.

Todd, A. S. (1959). The histological localisation of fibrinolysin activator. *J. Path. Bact.*, **78**, 281–283.

Toth, B. and Wilson, R. B. (1971). Blood vessel tumorigenesis by 1,2-dimethylhydrazine dihydrochloride (symmetrical) gross, light and electron microscopic descriptions. I. *Am. J. Path.*, **64**, 585–600.

Travers, B. (1844). *The Physiology of Inflammation and the Healing Process*. London.

Tromans, W. J., Horne, R. W., Gresham, G. A. and Bailey, A. J. (1963). Electron microscope studies on the structure of collagen fibrils by negative staining. *Z. Zellforsch. mikrosk. Anat.*, **58**, 798–802.

Ts'ao, C.-H. (1970). Graded endothelial injury of the rabbit aorta with special reference to platelet deposition. *Archs Path.*, **90**, 222–229.

Ts'ao, C.-H. and Glagov, S. (1970a). Basal endothelial attachment tenacity at cytoplasmic dense zones in the rabbit aorta. *Lab. Invest.*, **23**, 510–516.

Ts'ao, C.-H. and Glagov, S. (1970b). Platelet adhesion to subendothelial components in experimental aortic injury role of fine fibrils and basement membrane. *Brit. J. exp. Path.*, **51**, 423–427.

Ts'ao, C.-H., Glagov, S. and Kelsey, B. F. (1970). Special structural features of the rat portal vein. *Anat. Rec.*, **166**, 529–540.

Udenfriend, S. (1966). Formation of hydroxyproline in collagen. *Science*, **152**, 1335–1340.

Uehara, Y. and Burnstock, G. (1970). Demonstration of 'gap junctions' between smooth muscle cells. *J. Cell Biol.*, **44**, 215–217.

Uehara, Y., Campbell, G. R. and Burnstock, G. (1971). Cytoplasmic filaments in developing and adult vertebrate smooth muscle. *J. Cell Biol.*, **50**, 484–497.

Urschel, C. W., Covell, J. W., Sonnenblick, J. R., Jr and Braunwald, E. (1968). Effects of decreased aortic compliance on performance of the left ventricle. *Am. J. Physiol.*, **214**, 298–304.

Van Citters, R. L. (1966). Occlusion of lumina in small arterioles during vasoconstriction. *Circulation Res.*, **18**, 199–204.
Van Citters, R. L., Wagner, B. M. and Rushmer, R. F. (1962a). Architecture of small arteries during vasoconstriction. *Circulation Res.*, **10**, 668–675.
Van Citters, R. L., Wagner, B. M. and Rushmer, R. F. (1962b). Structural alterations in the arterial wall during vasoconstriction. *Cor. Vas*, **4**, 175–181.
Vane, J. R. (1969). The release and fate of vaso-active hormones in the circulation. *Brit. J. Pharmac. Chemother.*, **35**, 209–242.
Vastesaeger, M. M. and Delcourt, R. (1962). The natural history of atherosclerosis. *Circulation*, **26**, 841–855.
Vegge, T. (1971). An epithelial blood–aqueous barrier to horseradish peroxidase in the ciliary processes of the Vervet Monkey (*Cercopithecus aethiops*). *Z. Zellforsch. mikrosk. Anat.*, **114**, 309–320.
Veress, B., Bálint, A., Kóczé, A., Nagy, Z. and Jellinek, H. (1970). Increasing aortic permeability by atherogenic diet. *Atherosclerosis*, **11**, 369–371.
Verity, M. A. and Bevan, J. A. (1968). Fine structural study of the terminal effector plexus, neuromuscular and intermuscular relationships in the pulmonary artery. *J. Anat.*, **103**, 49–63.
Vogel, F. S. (1958). Enhanced susceptibility of proliferating endothelium to salivary gland virus under naturally occurring and experimental conditions. *Am. J. Path.*, **34**, 1069–1080.
Voigt, J. and Hansen, J. P. H. (1970). Spontaneous rupture of the aorta in young people, its relation to so-called medionecrosis cystica and Marfans syndrome – three medicolegal cases. *Acta path. microbiol. scand.*, A, **76**, Suppl. 212, 143–149.
Vost, A. (1969). Lipid accretion in the perfused rabbit aorta. *J. Atheroscler. Res.*, **9**, 221–238.
Voth, D., Schipp, R., Agsten, M., Schürmann, K., Kohlhardt, M. and Dudek, J. (1969). Untersuchungen über den Einfluss Kationenmilieus und verschiedener Pharmaka auf die Kontraktilität und Autorhythmik eines spontan aktiven glatten Gefässmuskels in vitro. *Arch. Kreislaufforsch.*, **60**, 364–387.
Wagenvoort, C. A. (1960). Vasoconstriction and medial hypertrophy in pulmonary hypertension. *Circulation*, **22**, 535–546.
Walford, R. L., Carter, P. K. and Schneider, R. B. (1964). Stability of labeled aortic elastic tissue with age and pregnancy in the rat. *Archs Path.*, **78**, 43–45.
Walker, F. W. and Attwood, H. D. (1960). The inferior vena cava opening in the diaphragm. *Brit. J. Surg.*, **48**, 86–88.
Warren, B. A. (1963). Fibrinolytic properties of vascular endothelium. *Brit. J. exp. Path.*, **44**, 365–372.
Warren, B. A. and Brock, L. G. (1964). The electron microscopic features and fibrinolytic properties of 'neo-intima'. *Brit. J. exp. Path.*, **45**, 612–617.
Warren, B. A. and De Bono, A. H. B. (1970). The ultrastructure of initial stages of platelet aggregation and adhesion to damaged vessel walls *in vivo*. *Brit. J. exp. Path.*, **51**, 415–422.
Wasserman, F. and Kubota, L. (1956). Observations on fibrillogenesis in the connective tissue of the chick embryo with the aid of silver impregnation. *J. biophys. biochem. Cytol.*, **2**, Suppl., 67–72.
Watson, M. L. (1958a). Staining of tissue sections for electron microscopy with heavy metals. *J. biophys. biochem. Cytol.*, **4**, 475–478.
Watson, M. L. (1958b). Staining of tissue sections for electron microscopy with heavy metals. II. Application of solutions containing lead and barium. *J. biophys. biochem. Cytol.*, **4**, 727–730.
Watts, H. F. (1961). Pathogenesis of human coronary artery atherosclerosis:

Demonstration of serum lipoproteins in the lesions and of localized intimal enzyme defects by histochemistry. *Circulation*, **24**, 1066 (Abstract).
Watts, S. H. (1907). The suture of blood vessels. Implantation and transplantation of vessels and organs. An historical and experimental study. *Johns Hopk. Hosp. Bull.*, **18**, 153–179.
Wearn, J. T., Mettier, S. R., Klumpp, T. G. and Zschiesche, L. J. (1933). The nature of the vascular communications between the coronary arteries and the chambers of the heart. *Am. Heart J.*, **9**, 143–164.
Webber, W. A. and Blackbourn, J. (1970). The permeability of the immature glomerulus to large molecules. *Lab. Invest.*, **23**, 1–7.
Weibel, E. R. and Elias, H. (1967). Quantitative methods in morphology. In *Proceedings of the Symposium on Quantitative Methods in morphology, 8th International Congress of Anatomists*. Springer-Verlag, Berlin, Heidelberg, N.Y.
Weibel, E. R., Kistler, G. S. and Scherle, W. F. (1966). Practical stereological methods for morphometric cytology. *J. Cell Biol.*, **30**, 23–38.
Weibel, E. R. and Palade, G. E. (1964). New cytoplasmic components in arterial endothelia. *J. Cell Biol.*, **23**, 101–112.
Weis-Fogh, T. and Andersen, S. O. (1970). New molecular model for the long-range elasticity of elastin. *Nature*, **227**, 718–721.
Weiss, J. M. (1953). The ergastoplasm – its fine structure and relation to protein synthesis as studied with the electron microscope in the pancreas of the Swiss albino mouse. *J. exp. Med.*, **98**, 607–618.
Weissmann, G. (1965a). Lysosomes. *New Engl. J. Med.*, **273**, 1084–1090.
Weissmann, G. (1965b). Lysosomes (concluded). *New Engl. J. Med.*, **273**, 1143–1149.
Weller, R. O. (1967). Cytochemistry of lipids in atherosclerosis. *J. Path. Bact.*, **94**, 171–182.
Wexler, B. C. (1964). Spontaneous arteriosclerosis in repeatedly bred male and female rats. *J. Atheroscler. Res.*, **4**, 57–80.
Wexler, B. C., Judd, J. T. and Kittinger, G. W. (1964). Changes in calcium, hydroxyproline and mucopolysaccharides in various aortic segments of arteriosclerotic breeder rats. *J. Atheroscler. Res.*, **4**, 397–415.
Whereat, A. F. (1967). Recent advances in experimental and molecular pathology. Atherosclerosis and metabolic disorder in the arterial wall. *Exp. and Mol. Path.*, **7**, 233–247.
Whiffen, J. D. and Gott, V. L. (1964). Effect of various surface active agents on heparin binding and clot formation on graphite surfaces. *Proc. Soc. exp. Biol. Med.*, **116**, 314–317.
White, J. F. (1954). Studies on the growth of blood vessels *in vitro*. I. The effect of initial pH on growth patterns. *Am. J. Anat.*, **94**, 127–169.
Widmann, J.-J., Cotran, R. S. and Fahimi, H. D. (1972). Mononuclear phagocytes (Kupffer cells) and endothelial cells identification of two functional cell types in rat liver sinusoids by endogenous peroxidase activity. *J. Cell Biol.*, **52**, 159–170.
Wiedeman, M. P. (1963). Dimensions of blood vessels from distributing artery to collecting vein. *Circulation Res.*, **12**, 375–378.
Wiederhielm, C. A. (1965). Distensibility characteristics of small blood vessels. *Fedn. Proc. Fedn. Am. Socs. exp. Biol.*, **24**, 1075–1084.
Wiederhielm, C. A., Woodbury, J. W., Kirk, S. and Rushmer, R. F. (1964). Pulsatile pressures in the microcirculation of frog's mesentery. *Am. J. Physiol.*, **207**, 173–176.
Wiener, J., Lattes, R. G. and Pearl, J. S. (1969). Vascular permeability and leukocyte emigration in allograft rejection. *Am. J. Path.*, **55**, 295–327.
Williams, G. (1957). A histological study of the connective tissue to Carrageenin. *J. Path. Bact.*, **73**, 557–563.

Williams, G. (1970). The pleural reaction to injury: a histological and electron-optical study with special reference to elastic-tissue formation. *J. Path. Bact.*, **100**, 1–7.

Williams, G. M., Krajewski, C. A., Dagher, F. J., Harr, A. M. ter, Roth, J. A. and Santos, G. W. (1971). Host repopulation of endothelium. *Transplantation Proc.*, **3**, 869–872.

Williams, R. G. (1944). Some properties of living thyroid cells and follicles. *Am. J. Anat.*, **75**, 95–119.

Williams, R. G. (1950). The microscopic structure and behaviour of spleen autografts in rabbits. *Am. J. Anat.*, **87**, 459–503.

Williams, R. G. (1951). The vascularity of normal and neoplastic grafts *in vivo*. *Cancer Res.*, **11**, 139–144.

Williams, R. G. (1954). Microscopic studies in living mammals with transparent chamber methods. *Int. Rev. Cytol.*, **3**, 359–398.

Williams, R. G. (1959). Experiments on the growth of blood vessels in thin tissue and in small autografts. *Anat. Rec.*, **133**, 465–485.

Williamson, J. R. and Grisham, J. W. (1961). Electron microscopy of leukocytic margination and emigration in acute inflammation in dog pancreas. *Am. J. Path.*, **39**, 239–256.

Williamson, J. R., Vogler, N. J. and Kilo, C. (1971). Regional variations in the width of the basement membrane of muscle capillaries in man and giraffe. *Am. J. Path.*, **63**, 359–370.

Willmer, E. N. (1945). Growth and form in tissue cultures. In *Essays on Growth and Form*, ed. W. E. L. Clark and P. B. Medawar. Clarendon Press, Oxford. pp. 264–294.

Winternitz, M. C. and Le Compte, P. M. (1940). Experimental infectious angiitis. *Am. J. Path.*, **16**, 1–12.

Winternitz, M. C., Thomas, R. M. and Le Compte, P. M. (1938). *The Biology of Arteriosclerosis*. Charles C. Thomas, Springfield, Illinois.

Wislocki, G. B. (1924). On the fate of carbon particles injected into the circulation with especial reference to the lungs. *Am. J. Anat.*, **32**, 423–445.

Wisse, E. (1970). An electron microscopic study of the fenestrated endothelial lining of rat liver sinusoids. *J. Ultrastruct. Res.*, **31**, 125–150.

Wissig, S. L. (1963). The anatomy of secretion in the follicular cells of the thyroid glands. II. The effect of acute thyrotrophic hormone stimulation on the secretory apparatus. *J. Cell Biol.*, **16**, 93–117.

Wissler, R. W. (1967). The arterial medial cell, smooth muscle, or multifunctional mesenchyme? (Editorial) *Circulation*, **36**, 1–4.

Wittenberg, J. B. and Wittenberg, B. A. (1961). Active transport of oxygen into the eye of fish. *Biol. Bull.*, **121**, 379 (Abstract).

Woerner, C. A. (1951). Microscopic anatomy of the arterial wall. *J. Geront.*, **6**, 165–166.

Wolff, J. (1963). Beiträge zur Ultrastruktur der Kapillaren in der normalen Grosshirnrinde. *Z. Zellforsch. mikrosk. Anat.*, **60**, 409–431.

Wolff, J. (1964). Ein Beitrag zur Ultrastruktur der Blutkapillaren: Das nahtlose Endothel. *Z. Zellforsch. mikrosk. Anat.*, **64**, 290–300.

Wolinsky, H. (1970). Response of the rat aortic media to hypertension. *Circulation Res.*, **26**, 507–522.

Wolinsky, H. (1971). Effects of hypertension and its reversal on the thoracic aorta of male and female rats. *Circulation Res.*, **28**, 622–637.

Wolinsky, H. and Glagov, S. (1964). Structural basis for the static mechanical properties of the aortic media. *Circulation Res.*, **14**, 400–413.

Wolinsky, H. and Glagov, S. (1967a). A lamellar unit of aortic medial structure and function in mammals. *Circulation Res.*, **20**, 99–111.

Wolinsky, H. and Glagov, S. (1967*b*). Nature of species differences in the medial distribution of aortic vasa vasorum in mammals. *Circulation Res.*, **20**, 409–421.

Wood, R. L. and Luft, J. H. (1965). The influence of buffer systems on fixation with osmium tetroxide. *J. Ultrastruct. Res.*, **12**, 22–45.

Woodard, W. C. and Pomerat, L. M. (1953). The development of patent blood vessels from adult human rib marrow in tissue culture. *Anat. Rec.*, **117**, 663–684.

Woodruff, C. E. (1926). Studies on the vasa vasorum. *Am. J. Path.*, **2**, 567–569.

Worthington, W. C. (1962). The blood vessels of the pituitary and the thyroid. In *Blood Vessels and Lymphatics*, ed. D. I. Abramson. Academic Press, N.Y. and London. pp. 428–464.

Wortman, B., Lee, K. T., Kim, D. N., Daoud, A. S. and Thomas, W. A. (1966). Studies related to protein synthesis in experimental animals fed atherogenic diets. II. DNA, RNA, and protein in aortas of rats fed atherogenic diets for one month. *Exp. and Mol. Path.*, Suppl. 3, 88–95.

Wright, I. (1963). The microscopical appearances of human peripheral arteries during growth and aging. *J. clin. Path.*, **16**, 499–522.

Yamada, E. (1955). The fine structure of the gall bladder epithelium of the mouse. *J. biophys. biochem. Cytol.*, **1**, 445–458.

Yoffey, J. M. and Courtice, F. C. (1970). *Lymphatics, Lymph and the Lymphomyeloid Complex*. Academic Press, N.Y. and London.

Zeigel, R. F. and Dalton, A. J. (1962). Speculations based on the morphology of the Golgi systems in several types of protein-secreting cells. *J. Cell Biol.*, **15**, 45–54.

Zelander, T., Ekholm, R. and Edlund, Y. (1962). The ultrastructural organisation of the rat exocrine pancreas. III. Intralobular vessels and nerves. *J. Ultrastruct. Res.*, **7**, 84–101.

Zellweger, J. P., Chapuis, G. and Mirkovitch, V. (1970). Conséquences morphologiques de l'emballage de l'aorte du chien dans une membrane de caoutchouc siliconé. Interruption expérimentale des vasa vasorum. *Virchows Arch. path. Anat. Physiol.*, **350**, 22–35.

Zemplényi, T. (1968). *Enzyme Biochemistry of the Arterial Wall as related to Atherosclerosis*. Lloyd–Luke, London.

Zugibe, F. T. (1963). Histochemical studies of human coronary atherogenesis: comparison with aortic and cerebral atherogenesis. *Circulation Res.*, **13**, 409.

Zweifach, B. W. (1937). The structure and reactions of the small blood vessels in amphibia. *Am. J. Anat.*, **60**, 473–514.

Zweifach, B. W. (1971*a*). Local regulation of capillary pressure. *Circulation Res.*, **28**, Suppl. 1, 129–134.

Zweifach, B. W. (1971*b*). 1971 E. M. Landis Award acceptance speech. *Microvasc. Res.*, **3**, 345–353.

AUTHOR INDEX

Aars, H. 11, 86
Aarseth, P. 26
Abraham, A. 132, 133, 136, 137, 138, 139
Ackerman, G. A. 47
Adams, C. W. M. 108, 126, 129, 131
Adelson, E. 39, 65
Agsten, M. 20, 76
Aikawa, M. 90, 91
Akester, A. R. 24, 25
Akmayev, I. G. 24
Albala, A. 50
Albert, E. N. 117, 120
Alex, M. 111, 112, 113
Alexander, A. F. 76, 118
Algire, G. H. 63
Allison, A. C. 43, 94
Allison, F., Jr 150, 152
Altura, B. M. 85, 153
Ancla, M. 74, 136, 154
Andersen, S. O. 110
Andrew, W. 67
Angevine, D. M. 109
Anitschkow, N. N. 13, 108
Arey, L. B. 69, 120
Aristizabal, N. 108
Arndt, J. O. 13, 86
Aschoff, L. 90, 106
Ashford, T. P. 44
Astrup, T. 43, 44
Attwood, H. D. 6, 91
Austin, J. 38

Babala, J. 53
Bagley, E. H. 14, 69, 71
Bahr, G. F. 55
Bailey, A. J. 111
Bainton, D. F. 50
Balboa, R. S. 38

Baldwin, D. 90, 93, 118
Bálint, A. 52, 130, 131, 132
Balis, J. U. 72, 78, 80
Balis, J. V. 113
Baló, J. 112
Banfield, W. G. 60
Banga, I. 112
Baradi, A. F. 48
Barber, V. C. 136
Barer, R. 55
Barker, B. E. 31, 53
Barnard, W. G. 65
Barquet, J. 103
Barr, L. 84
Barr, M. L. 61
Barrnett, R. J. 40, 49, 55
Barton, S. 5
Batson, O. V. 18
Bauer, H. 126
Baumgartner, H. R. 106
Bayliss, Olga B. 108, 126, 129, 131
Bazett, H. C. 20
Beard, J. W. 26, 52
Beard, L. A. 26, 52
Becker, A. E. 91
Becker, C. G. 81, 90, 155
Behnke, O. 32
Bell, C. 136, 139
Bencosme, S. A. 113
Benditt, E. P. 72, 74, 87, 88, 91, 92, 104, 115
Bennett, H. S. 27, 31, 33, 55, 97, 99
Bennett, T. 7, 24, 137
Bensch, K. 49, 55
Benson, I. C. 107
Bentley, J. P. 87, 110
Berg, H. F. 87, 109
Bergel, D. H. 105

Berlepsch, K. von 33, 109
Berliner, R. W. 60, 72
Berne, R. M. 5
Berry, C. L. 90
Bertram, L. F. 61
Bevan, J. A. 133, 135, 138
Bhadrakom, S. 85
Bhisey, A. N. 43
Bhussry, B. R. 117, 120
Biempica, L. 50
Billroth, T. 46
Bing, R. J. 64
Biological Handbooks 2
Birch, C. L. 18, 24, 61
Björkerud, S. 109
Björnheden, T. 63, 89
Blache, J. O. 120, 121
Blackbourn, J. 57, 59
Bladen, H. A. 121
Blake, T. M. 130
Blanchette-Mackie, E. J. 48, 59
Bland, E. F. 140
Bloch, E. H. 23, 26
Bloom, F. E. 133
Bloom, S. 55
Bloom, W. 87, 120
Blumenthal, H. T. 120, 121
Boak, J. L. 28
Böck, P. L. 60
Bodenheimer, T. S. 54, 101
Bodian, D. 133, 135
Bohr, D. F. 85, 86
Bolton, T. B. 83
Bond, T. P. 15
Bond, V. P. 67, 96
Bondjers, G. 63, 89
Bornstein, P. 113, 114

Author index

Borrero, L. M. 54, 101
Bossert, W. H. 51
Bourdeau, J. E. 59
Bouzarth, W. F. 5
Boyd, W. 26
Braasch, D. 15
Brash, J. C. 22, 28
Braunwald, E. 107
Breen, P. C. 26
Brenner, R. M. 46, 60
Brightman, M. W. 54, 101
Brindley, D. C. 60
Brine, K. L. 150, 152
Brock, L. G. 63, 120
Brookes, M. 23, 70
Brown, B. A. 26, 27
Bruns, R. R. 33, 35, 48, 50, 51, 53, 54, 57, 60, 61, 68, 93
Brux, J. de 74, 136, 154
Buck, R. C. 31, 48, 52, 77, 87, 90, 130
Buckley, I. K. 152, 153
Buluk, K. 43
Bunting, C. H. 120
Burch, G. E. 66
Burkel, W. E. 23, 69, 74, 136
Burn, J. H. 133
Burns, W. F. 126
Burnstock, G. 48, 78, 82, 84, 86, 94, 133, 135, 136, 137, 138, 139
Burri, P. H. 43
Burrows, L. 64
Burton, A. C. 40, 41, 77, 83, 104, 105, 106, 117, 118, 119, 120, 145, 146
Byers, S. O. 93, 120

Caen, J. P. 106
Caesar, R. 82
Camp, P. D. 140
Campbell, G. R. 82, 94
Campbell, W. G., Jr 91
Cappell, D. F. 26
Caro, L. G. 11, 44, 86
Carone, F. A. 59
Carter, D. 126
Carter, P. K. 120
Casanova, S. 59
Casley-Smith, J. R. 48, 51
Castellani, A. 111, 112

Catchpole, H. R. 97, 99
Caulfield, J. B. 55
Cauna, N. 76, 77, 117
Cavallo, T. 64
Chakmakjian, S. 6, 86, 139
Chalkey, H. W. 63
Chambers, R. 32, 43, 69, 71, 74
Chan, A. S. 72, 78, 80
Chang, J. J. 28
Chapuis, G. 123, 126
Cheville, N. F. 40, 66
Chiba, C. 64
Chick, W. J. 133, 136
Chodkowski, K. 63
Christie, G. H. 28
Chrysohou, A. 64
Claesson, M. H. 99, 155
Clark, E. L. 31, 43, 63, 64, 68, 69, 70, 75, 133, 150, 152, 154
Clark, E. R. 28, 31, 43, 63, 64, 68, 69, 70, 75, 133, 150, 152, 154
Clark, W. G. 40
Clarke, J. A. 127, 128, 129
Claude, A. 44
Claxton, E. B. 41
Cleary, E. G. 53, 93, 110, 111, 112, 114
Cliff, W. J. 35, 37, 38, 43, 44, 46, 47, 52, 61, 63, 65, 66, 67, 68, 69, 70, 72, 77, 78, 80, 81, 82, 83, 84, 86, 87, 88, 91, 92, 94, 98, 104, 106, 107, 109, 111, 112, 113, 114, 115, 119, 121, 150, 152, 154
Coccheri, S. 44
Cogan, D. G. 66, 68, 70, 71
Cohn, Z. A. 94
Collet, A. 4, 37, 46, 77, 78, 82, 97, 98
Comfort, A. 121
Comroe, J. H., Jr 138
Conen, P. E. 72, 78, 80
Constantinides, P. 40, 154
Cook, M. L. 89
Cooke, P. H. 55, 82, 90
Cookson, F. B. 77, 87, 96
Cooper, J. H. 113

Copley, A. L. 33
Corner, G. W. 103
Cotran, R. S. 26, 27, 43, 50, 52, 61, 63, 64, 68, 71, 96, 98, 99, 154
Cottrell, G. A. 1
Coulston, F. 90, 93
Coupland, R. E. 133
Courtice, F. C. 2, 3, 93, 130, 131, 153
Covell, J. W. 107
Cowdry, E. V. 131
Cox, R. C. 110, 111, 112
Cox, R. W. 103
Crane, W. A. J. 63, 90
Crawford, T. 89, 106, 109
Crocker, D. J. 64, 68, 69, 70
Cromwell, S. B. 87, 88
Cruikshank, B. 102
Curran, R. C. 33
Curtis, A. S. G. 32, 42, 55

Dagher, F. J. 61, 63
Dahl, L. K. 67, 96
Dale, H. H. 153
Dalhamn, T. 111
Dalton, A. J. 44, 86, 93, 94
Dann, L. 63
Daoud, A. 87, 90, 93, 108
Daoud, A. S. 89, 90, 93
Davidson, D. G. 60, 72
Davies, P. 43, 94
Davis, H. F. 110, 112
Davison, A. N. 129, 131
Day, R. 20
De Bakey, M. E. 63
De Bono, A. H. B. 106
De Bruyn, P. P. H. 26
Delcourt, R. 90, 108
Dempsey, E. W. 111, 112, 113, 121
Dennison, S. M. 87, 90
De Petris, S. 43, 94
De Robertis, E. 112
Derrick, J. R. 15
Dessel, B. H. 38
Dettmer, N. 111, 112
Devine, C. E. 81, 89, 133, 136, 137, 138
Devis, R. 92
Dewey, M. M. 84

Author index

Dickinson, C. J. 23
Dintenfass, L. 9
Dixon, F. J. 102
Dobb, M. G. 111
Dobson, E. L. 26, 27
Dorfman, A. 110
Doty, P. 103, 111
Dreiling, D. 64
Duchacek, H. 13
Dudek, J. 20, 76
Duff, G. L. 77, 129
Duguid, J. B. 44, 106, 107, 109
Duling, B. R. 5
Dunihue, F. H. 84, 89, 102, 103, 130
Dunphy, J. E. 63
Du Shane, J. W. 115
Dustin, A. P. 63
Dutta, L. P. 63, 90
Duve, C. de 52, 94

Easty, G. C. 33
Ebel, A. 106
Ebert, R. H. 69, 96, 150, 152
Eden, M. 60, 72
Edlund, Y. 57, 60, 74, 80, 82
Edwards, G. A. 82
Edwards, J. E. 115
Edwards, L. C. 63
Edwards, R. H. 64
Ehinger, B. 26, 136, 137
Einheber, A. 126
Eisenberg, L. 20
Eisenberg, S. 72, 78, 87
Ekholm, R. 57, 60, 74, 80, 82
Elias, H. 87
Elliott, D. F. 153
Ellis, H. D. 11
Ellison, G. D. 5, 139
Elsden, D. F. 110
Elsner, R. 6
Esterly, J. R. 66
Estes, P. C. 40, 66
Evans, C. L. 13, 17, 22, 29

Fahimi, H. D. 26, 27, 50
Fahrenbach, W. H. 53, 93, 111, 112, 114

Falck, B. 26, 133, 134, 136, 137
Fanger, H. 53
Fani, K. 90, 93, 108
Farner, D. S. 24, 37, 38, 60, 68
Farnes, P. 31, 53
Farquhar, M. G. 34, 50, 55, 59, 68, 69, 70, 72, 99, 102
Farrant, J. L. 49, 50
Farrell, G. A. 11
Fawcett, D. W. 31, 32, 35, 37, 44, 46, 56, 57, 60, 68, 69, 71, 72, 86, 97, 98
Fay, S. F. 55, 82, 90
Fedorko, M. E. 48
Feeney, L. 74, 76, 82, 98, 117, 136
Fekety, F. R. 152
Feldman, J. D. 101
Feldman, S. A. 105, 121
Felix, M. D. 93, 94
Fennessy, J. F. 39, 65
Fernando, N. V. P. 61, 91, 154
Ferreira, J. F. D. 37, 48
Ferrer, J. 24
Fillenz, M. 26, 133, 135, 136, 138
Filo, R. S. 85
First Conference on Microcirculatory Physiology and Pathology 147
Fitz-Gerald, J. M. 11
Florentin, R. 90, 93, 108
Florey, Lord 31, 48, 50, 51, 55, 56, 60, 99, 120
Florey, H. W. 31, 32, 48, 61, 63, 69, 90, 96, 98
Folkman, J. 43, 63, 64
Folkow, B. 18, 90
Fontaine, R. 106
Foot, N. C. 52, 63, 64
Foreman, J. E. K. 13
Forster, R. 20
Fowler, N. O. 4
Franchi, C. M. 112
Franke, W. W. 78
Franklin, K. J. 4, 6, 18, 19, 76, 91, 105, 115, 118, 128, 136, 140
Fraser, R. D. B. 102, 111

Frayser, R. 68, 148
Freed, J. J. 43
Freiman, D. G. 44
French, J. E. 31, 46, 48, 52, 77, 90, 91, 98, 103, 105, 154
Friedman, M. 93, 120
Fritz, K. E. 87
Fujinami, T. 106
Fulton, G. P. 69, 71, 85, 139
Furchgott, R. F. 85
Furness, J. B. 137
Furnival, C. M. 4, 25
Furth, J. 53

Gabbiani, G. 92
Gabella, G. 78
Gannon, B. 84, 136, 137, 138, 139
Ganote, C. E. 59
Garamvölgyi, N. 81
Garbarsch, C. 126
Gartley, K. M. 109
Gautvik, K. 5
Geer, J. C. 64, 68, 69, 70, 90, 108
Geiringer, E. 128, 130
Gerrity, R. G. 35, 37, 38, 43, 44, 46, 52, 61, 66, 77, 80, 83, 87, 93, 98, 103, 104, 106, 109, 113, 114, 115, 120, 154
Gersh, I. 97, 99
Ghidoni, J. J. 87, 90, 93, 107
Giacomelli, F. 33, 54, 101, 154
Gillman, T. 109, 123
Giltaire-Ralyte, L. 77, 78, 82
Gimbrone, M. A., Jr 43, 63
Girling, F. 41
Gisslen, J. L. 137, 139
Glagov, S. 10, 11, 24, 43, 72, 76, 98, 102, 104, 105, 106, 115, 118, 119, 120, 121, 125, 137, 142
Glasser, O. 40, 41
Gold, L. D. 6
Goldfischer, S. 50
Goldsmith, H. L. 9, 15

Author index

Goodenough, D. A. 55
Goodman, P. 46
Gore, I. 38, 57, 60, 106
Gott, V. L. 41
Gotte, L. 110, 111, 112
Gowans, J. L. 155
Graf, J. 40
Grafflin, A. L. 14, 69, 71
Graham, J. D. P. 133, 135, 136, 137
Graham, R. C. 150, 152
Graham, R. C., Jr 49, 50, 59, 101
Grant, L. 103, 150, 152
Grant, R. A. 103, 104, 115
Grant, R. T. 22, 140
Gray, S. H. 120, 121
Graziadei, P. 136
Green, H. D. 3, 4, 6, 8, 9, 10, 11, 17
Greenlee, T. K. 113, 114
Greenwalt, T. J. 38
Greenway, C. V. 23, 27, 59, 153
Greer, S. J. 63
Gregg, D. E. 13, 77, 129
Gresham, G. A. 89, 108, 111
Grisham, J. W. 37
Groat, R. A. 9
Gross, J. 103, 111, 112
Grotte, G. 54
Grubb, D. J. 26, 27, 59, 68, 101
Gudbjarnason, S. 64
Guest, M. M. 15
Guntheroth, W. G. 6, 86, 139
Guthe, K. F. 85
Gutstein, W. H. 10, 11, 90, 109
Guttman, R. D. 64
Guttuta, Monika L. 52, 68, 99

Haar, J. L. 47
Hadfield, G. 69, 98
Haglund, U. 18
Haimovici, H. 109
Hall, C. E. 103, 111
Hall, D. A. 110, 111, 112, 123

Hallbäck, M. 90, 91
Hallock, P. 107
Halpert, B. 63
Ham, K. N. 154
Hama, K. 20, 68, 71, 97, 154
Hammock, M. K. 60
Hammond, D. D. 6
Hammond, G. L. 29
Hampton, J. C. 27, 31, 97
Hanafee, W. N. 6
Handler, F. P. 120, 121
Hansen, J. P. H. 124
Hanson, A. N. 87, 110
Hanson, J. 81
Harb, J. M. 66
Hardaway, R. M. 41, 44
Harker, L. A. 38
Harms, R. H. 78, 91, 111, 112, 124
Harr, A. M. ter 61, 63
Harris, H. 150, 152
Harris, R. 127, 128
Hartman, J. L. 113, 114
Hass, G. M. 64, 83, 89, 90, 93, 96, 107, 108, 109, 113, 114, 115, 117, 118, 119, 120, 121, 124
Hassler, O. 77, 93, 96, 106, 118
Hathorn, M. 109, 123
Haust, M. D. 72, 82, 91, 113, 114
Haynes, F. 115, 118
Heath, D. 115
Heitzman, E. J. 39, 65
Helfer, H. 20, 24
Helin, P. 126
Heller, A. L. 4
Heron, I. 48, 64
Higginbotham, A. C. 125, 128, 129, 130, 131
Higginbotham, F. H. 125, 128, 129, 130, 131
Hilker, D. M. 120
Hill, A. G. S. 102
Hill, C. H. 124
Hilton, S. M. 86
Hirsch, J. G. 48
Ho, P.-L. 110
Hock, R. 20
Hodge, A. J. 56
Hoff, H. F. 40

Hogan, M. J. 74, 76, 82, 98, 117, 136
Holfreter, J. 43
Hollander, W. 108
Holman, M. E. 135
Honig, C. R. 81
Hope, J. 48
Hopwood, D. 133
Horne, R. W. 103, 111
Horowitz, R. E. 64
Horton, E. W. 153
Howard, A. N. 89, 108
Howard, J. G. 28
Hughes, A. F. W. 63
Hume, D. M. 48, 64
Hunter, D. 4
Hurley, J. V. 98, 150, 152, 154
Hutchison, K. J. 13
Hüttner, I. 48, 50, 52, 53, 96
Huxley, H. E. 81

Ibrahim, M. Z. M. 108, 126, 129, 131
Ikeda, M. 26, 133, 136, 137
Illingworth, C. F. W. 2
Imai, H. 77, 90
Intaglietta, M. 8, 10
Irey, N. S. 38
Irle, C. 92
Irvine, G. 133, 136
Irving, L. 20
Ishii, S. 84
Iwayama, T. 84, 136, 137, 138, 139

Jackson, D. S. 102, 110, 114
Jackson, S. 102
Jackson, S. F. 92, 103
Jacobs, E. 40
Jacobson, W. 39, 92, 97, 102, 103
James, D. W. 92
Jamieson, E. B. 22, 28
Jaques, R. 20, 24
Jarmolych, J. 87, 90, 93, 108
Jellinek, H. 52, 130, 131, 132
Jennings, M. A. 48, 50, 51, 55, 56, 90, 99, 120

Author index

Jensen, R. 76, 118
Jensen, R. E. 13
Jepson, P. 115
Jerrard, W. 41
Jodal, M. 18
Johansen, K. 2, 3, 4, 13, 20, 115
Johansson, B. 5, 6, 13, 20, 86, 139
Johnson, R. A. 130, 131, 132
Johnson, S. A. 38
Johnson, W. W. 38
Jones, A. L. 26, 27, 59, 68, 101
Jones, R. 90, 93
Joó, F. 53
Jordan, G. L., Jr 63
Jorgensen, L. 131
Jorgensen, O. 99, 155
Joseph, S. 55
Judd, J. T. 109

Kabat, E. A. 53
Kadis, B. 46
Kallman, F. 55
Kampmeier, O. F. 18, 24, 61
Kao, K.-Y. T. 106, 120
Kao, V. C. Y. 81, 90, 155
Kaplan, D. 109
Karatzas, N. B. 4, 14
Kark, A. F. 64
Karlin, L. J. 57, 59
Karlsbad, G. 94
Karnovsky, M. J. 33, 49, 50, 51, 54, 56, 57, 59, 61, 99, 101
Karpinski, E. 5
Karrer, H. E. 20, 57, 72, 78, 81, 91, 94, 104, 106, 107
Kasai, T. 63, 87
Katz, A. I. 6
Katz, L. N. 130
Keech, M. K. 72, 82, 115, 119, 142
Kelly, F. B., Jr 90
Kelly, L. S. 26, 27
Kelly, R. E. 81
Kelsey, B. F. 24, 76, 118, 137
Kempf, E. 106

Kershaw, T. 74, 82, 105
Keyserlingk, D. Graf 81, 82
Kiil, F. 144
Kilo, C. 102
Kim, D. N. 89, 90
King, A. J. 89, 108
King, A. L. 112, 121
Kirby-Smith, H. T. 28
Kirk, J. E. 66, 125
Kirk, S. 14
Kisch, B. 35, 37
Kiser, J. 63
Kistler, G. S. 87
Kittinger, G. W. 109
Klebanoff, S. J. 86, 88
Klumpp, T. G. 30
Knieriem, H. J. 81, 90, 155
Knight, E. J. 155
Knisely, M. H. 29
Knoll, J. 81
Kóczé, A. 52
Kohlardt, M. 20, 76
Koletsky, S. 90, 91
Kontos, H. A. 5
Korner, P. I. 138, 140
Kraemer, D. M. 120
Krajewski, C. A. 61, 63
Kramer, H. 102
Krogh, A. 4, 5, 8, 14, 17, 18, 43, 68, 137, 146, 150, 153, 154
Kruml, J. 109
Kubota, L. 102
Kuron, G. W. 123
Kurtz, S. M. 101
Kuwabara, T. 66, 68, 70, 71

Ladányi, P. 89
Lalich, J. L. 109
Lancaster, M. G. 152
Landis, E. M. 14, 17, 54, 60, 61, 68, 140, 148, 150, 152
Lane, B. P. 84
Lansing, A. I. 109, 110, 111, 112, 113, 120, 121, 123
Laschi, R. 59
La Taillade, J. N. 90, 109

Latta, H. 50, 56, 57, 59, 89, 97, 99, 101
Latta, J. S. 46
Lattes, R. G. 48
Laursen, T. J. S. 66, 125
Lautsch, E. V. 77
Lawton, R. W. 112, 121
Lazzarini-Robertson, A., Jr 90, 109
Leak, L. V. 37, 55, 97
Learoyd, B. M. 13, 106, 107
Le Compte, P. M. 125, 127, 128, 129, 130
Lee, G. de J. 10
Lee, K. J. 77, 90
Lee, K. T. 77, 89, 90
Lee, S. K. 77, 90
Leeson, T. S. 56, 57
Legrand, Y. 106
Lelkes, Gy. 89
Lesko, A. 53
Levene, C. I. 89, 105, 106, 109
Leventhal, M. J. 61, 154, 155
Lever, J. D. 133, 135, 136, 137
Levick, J. R. 152
Levinsky, N. G. 60, 72
Lewis, G. P. 5, 153
Lewis, L. 10, 11
Lewis, M. S. 103
Lewis, R. N. 4
Lewis, T. 140, 150, 152, 153
Lewis, W. H. 47, 94
Lie, M. 144
Light, L. H. 13
Lillywhite, J. W. 63
Linden, R. J. 4, 25
Lindquist, R. R. 64
Lindsay, H. A. 61
Linhart, J. 109
Little, K. 102, 110, 111, 112
Ljung, B. 20, 84, 138, 139
Loeven, W. A. 123
Logan, M. A. 104
Lojda, Z. 53
Longley, J. B. 60
Lorenzen, I. 126
Love, L. 20

Author index

Lowy, J. 81
Ludatscher, R. M. 5, 20, 37, 57, 65, 68, 98
Luft, J. H. 27, 31, 32, 55, 57, 97, 116, 154
Lund, B. 40, 48, 64
Lundgren, O. 18
Lundgren, Y. 90, 91
Lundvall, J. 5
Lutz, B. R. 69, 71, 85, 139

McCombs, H. L. 108
McConnell, J. G. 20, 76
McCuskey, R. S. 23, 26, 69
McDonald, D. A. 11, 13
McGandy, R. B. 108
McGavack, T. H. 106, 120
McGill, H. C., Jr 90, 108
McIlroy, M. B. 4, 14
McIntosh, G. H. 28
McKay, D. 41, 44
McKay, D. G. 38, 66
McMillan, G. C. 77
McPhaul, J. J., Jr 102
Mack, G. 106
Mackaness, G. B. 94
Macrae, T. P. 111
Maier, N. 109
Majno, G. 26, 27, 31, 33, 50, 52, 61, 68, 69, 70, 92, 97, 98, 99, 154, 155
Malmfors, T. 7, 24, 137
Mancini, R. E. 103
Manion, W. C. 38, 40
Manly, R. S. 40
Mann, S. P. 24
Marbach, E. P. 102
Marchesi, V. T. 38, 40, 48, 61, 98, 150, 154
Margaretten, W. 38, 66
Mark, J. S. T. 88
Marshall, J. R. 96
Martin, A. V. W. 102
Martin, A. W. 2, 3, 4, 13, 20, 115
Martin, G. R. 121
Matsubara, F. 32
Matthews, J. 5
Matthiessen, M. E. 126
Maunsbach, A. 55

Maunsbach, A. B. 89
Maximov, A. 31
Maximow, A. 102
Mayerson, H. S. 27, 53
Meadors, M. P., Jr 38
Meek, G. A. 31, 32, 55
Mellander, S. 5, 6, 13, 20, 86, 139
Meneghelli, V. 111, 112
Menkin, V. 152
Menzies, D. W. 121
Mercer, E. H. 33
Merkow, L. 109
Merril, J. P. 64
Merrillees, N. C. R. 48, 78, 133, 135, 136, 137, 138
Mettier, S. R. 30
Meyer, E. 29
Meyer, J. 130
Meyer, K. 109
Michaelis, A. R. 3
Michel, C. C. 152
Michels, N. A. 23
Mikami, S.-I. 24, 37, 38, 60, 68
Mikata, A. 50, 51, 52, 155
Milhorat, T. H. 60
Miller, H. 120
Mims, C. A. 66
Mirkovitch, V. 123, 126
Mittler, B. S. 55
Moffat, D. B. 60, 72, 135
Moggio, R. A. 29
Molinari, S. 74, 78, 81, 82, 86, 87, 105, 117
Monie, I. W. 77
Monto, R. W. 38
Moon, V. H. 152, 154
Moore, D. H. 48, 50, 74
Moore, J. 137
Moore, R. D. 91, 117
Morato, M. J. X. 37, 48
More, R. H. 48, 50, 52, 53, 91, 96, 113
Moreno, A. H. 6
Moritz, A. R. 89, 93, 117
Morris, B. 28, 48
Morris, H. P. 63
Moses, J. M. 150, 152
Mosher, M. B. 60
Moss, N. S. 72, 74, 91, 104, 115

Movat, H. Z. 61, 91, 154
Muir, A. R. 33
Munger, B. L. 44
Murad, T. M. 64, 68, 69, 70
Murphy, C. F. 60
Murphy, G. E. 81, 90, 155
Murray, M. 87, 109
Mustard, J. F. 131
Myhre Jensen, O. 40, 48, 64

Nagasawa, S. 63, 87
Nagy, Z. 52, 130, 131, 132
Neuman, R. E. 104
Newton, M. 20
Nichol, J. 41
Nickerson, N. D. 4
Nicoll, P. A. 19, 68, 69, 74, 75, 76, 85, 148
Niki, R. 50, 51, 52, 155
Nishihara, H. 83
Nishimaru, Y. 1, 2
Noble, M. I. M. 4, 14
North, A. C. T. 102
North, R. J. 40, 94
Northover, A. M. 153, 154
Northover, B. J. 153, 154
Nour-Eldin, F. 38
Novikoff, A. B. 50, 52, 53
Nylen, M. 121
Nyström, S. H. M. 116, 117, 123

Oberling, C. 44, 48
O'Dell, B. L. 110
Odland, G. F. 87, 90
Ohta, G. 32
Oksche, A. 24, 37, 38, 60, 68
Olsen, B. R. 103
O'Neal, R. M. 63, 77, 87, 90, 93, 96, 107
Oosaki, T. 84
Opdyke, D. F. 29
Oppenheimer, E. H. 66
Ott, W. H. 123

Packham, M. A. 131
Palade, G. E. 33, 34, 35, 43, 46, 48, 50, 51, 52, 53, 54, 55, 57, 59, 60, 61, 68, 69, 70, 72, 86, 92, 93, 94, 97, 98, 99, 102, 154

Author index

Palay, S. L. 57, 59
Palmer, P. 103, 150, 152
Panner, B. J. 81
Pantesco, V. 106
Pappas, G. D. 48, 50, 56, 57, 59, 97, 101, 154
Pappenheimer, J. R. 54, 61, 68, 101, 148
Parker, F. 83, 87, 90, 98, 113, 116
Parnell, J. 13
Paronetto, F. 64
Parpart, A. K. 28
Partridge, S. M. 110, 111, 112, 113, 120, 121
Paterson, J. C. 109
Patterson, J. L., Jr 5
Paule, W. J. 72, 82, 83, 97, 98, 102, 104, 115, 119
Payling Wright, H. 22
Paz, M. A. 103
Peachey, L. D. 48
Pearl, J. S. 48
Pearse, A. G. E. 113
Pearson, B. 64
Pease, D. C. 72, 74, 78, 81, 82, 83, 86, 87, 97, 98, 102, 105, 115, 117, 119
Pedersen, N. C. 48
Penn, J. 109, 123
Pentreath, V. W. 1
Peracchia, C. 55
Pernis, B. 94
Peters, A. 33
Peterson, L. H. 13, 104, 106
Philps, P. C. 116, 154
Picken, L. E. R. 103
Piez, K. A. 103
Policard, A. 4, 37, 46, 77, 78, 82, 97, 98
Pollak, O. J. 63, 87
Pomerat, L. M. 31, 48, 63
Poole, J. C. F. 31, 32, 61, 63, 87, 88, 90, 96, 105, 106
Porter, K. R. 44, 46, 55, 92
Pregermain, S. 4, 46, 97, 98
Přerovský, I. 109
Prince, L. M. 41
Prosser, C. L. 1, 3

Quick, A. J. 38, 41, 64, 65

Rabin, E. R. 63
Radawski, D. 5
Ramos, H. 64
Ramsey, E. M. 29, 117
Randall, J. T. 102
Raper, A. J. 5
Ratnoff, O. D. 152
Reddy, R. V. 6
Reed, R. 110, 111, 112
Rees, P. M. 115
Renkin, E. M. 54, 101
Reuben, S. R. 10
Revel, J. P. 55
Rex, R. O. 28
Rhodin, J. A. G. 35, 44, 46, 48, 56, 57, 60, 68, 69, 70, 72, 74, 75, 77, 78, 82, 83, 84, 91, 92, 97, 103, 105, 111, 113, 118, 136, 137, 138, 145
Rice, R. V. 81
Rivkin, L. M. 120
Roach, M. R. 77, 119, 123, 124
Robbins, E. 55
Robb-Smith, A. H. T. 65
Roberts, J. T. 121
Robertson, C. E. 98, 102
Robertson, H. F. 127, 128, 129, 132
Robertson, J. D. 33, 41, 56, 82
Robinson, D. S. 90
Robinson, M. 40, 154
Rocha, H. 152
Rodbard, S. 4, 5, 9
Roddie, I. C. 20, 76
Rolewicz, T. F. 137, 139
Romer, A. S. 3, 4, 22, 23, 24
Rona, G. 48, 50, 52, 53, 96
Ropke, C. 99, 155
Rosen, H. 126
Rosenblum, W. I. 9, 17
Rosenbluth, J. 53
Rosenthal, T. B. 111, 112, 113
Ross, R. 63, 86, 87, 88, 91, 92, 113, 114

Roth, J. A. 61, 63
Rothbard, S. 97
Rowley, D. A. 154
Rowsell, H. C. 131
Rüegg, J. C. 81
Rushmer, R. F. 14, 144
Ruska, H. 48, 50, 74, 82
Ryan, G. B. 92, 153, 154

Sabatini, D. D. 49, 55
Sabin, F. R. 47
Sade, R. 64
Sainte-Marie, G. 155
St George, S. 93
Salgado, E. D. 90, 91
Samuelson, A. 91
Sandberg, L. B. 53, 93, 111, 112, 114
Sanders, A. G. 31, 69, 103, 150, 152
Sanderson, M. H. 99
Sandison, J. C. 28, 31, 68, 69, 147
Santos, G. W. 61, 63
Santos-Buch, C. A. 91
Sarmenta, S. S. 64
Sasaki, H. 32
Saunders, K. B. 4, 14
Schachter, M. 5
Schallock, G. 117, 118
Schaper, W. 63
Scharfstein, H. 10, 11
Scheinthal, B. M. 33
Schenck, D. J. 11
Scherle, W. F. 87
Schevill, W. E. 20
Schiffman, E. 121
Schipp, R. 20, 76
Schlichter, J. 127, 128
Schlichter, J. G. 130
Schloss, G. 93
Schmidt-Diedrichs, A. 93
Schmitt, F. O. 103
Schneider, R. B. 120
Schneider, W. C. 48
Schoefl, G. I. 46, 48, 52, 61, 70, 98, 155
Schoenberg, M. D. 91, 117
Scholander, P. F. 20
Schrodt, G. R. 87, 109
Schroter, R. C. 11
Schulman, J. H. 41

Schumacker, H. B., Jr 93
Schürmann, K. 20, 76
Schwarz, W. 111, 112
Scott, J. B. 5
Scott, R. C. 4
Scott, R. F. 90, 93
Scow, R. O. 48, 59
Secker Walker, R. H. 23
Seifert, K. 77
Seitner, M. M. 107
Sejersted, O. M. 144
Selkurt, E. E. 29
Sengel, A. 43, 44
Serafini-Fracassini, A. 110, 111, 112
Sethi, N. 70
Shea, S. M. 51, 61, 154, 155
Shenkin, H. A. 5
Shepherd, W. E. 98, 102
Shirahama, T. 106
Shirley, H. H. 27, 53
Short, D. 116, 117
Siegel, H. 123
Siegesmund, K. A. 38
Siggins, G. R. 133
Siller, W. G. 115
Silva, D. G. 26, 133, 136, 137
Simmonds, W. J. 3
Simons, J. R. 3
Simpson, C. F. 78, 91, 111, 112, 124
Simpson, F. O. 89, 133, 136, 137, 138
Slichter, S. J. 38
Slinger, S. J. 109
Small, J. V. 81
Smith, C. 107
Smith, D. E. 49
Smith, J. B. 28
Smith, J. W. 18, 127
Smith, Mary R. 150, 152
Smuckler, E. A. 101
Snodgrass, M. J. 26, 29
Snook, T. 26, 29
Snow, H. M. 4, 25
Sohal, R. S. 66
Solberg, L. A. 11
Somlyo, A. P. 20, 78, 81, 85, 86, 136, 139
Somlyo, A. V. 20, 78, 81, 85, 86, 136, 139

Sonnenblick, J. R., Jr 107
Sorokin, S. 80, 86
Spaet, T. H. 106
Spain, D. M. 108
Spector, W. G. 153
Spiro, D. 33, 54, 101, 154
Sporrong, B. 26, 136, 137
Spraragen, S. C. 67, 96
Spriggs, T. L. B. 133, 135, 136, 137
Städeli, H. 104, 107
Staehelin, L. A. 37. 56
Stark, R. D. 23, 27, 59, 153
Starling, E. H. 148
Statkov, P. R. 92
Stearner, S. P. 99
Stegall, H. F. 13, 86
Stehbens, W. E. 5, 11, 12, 29, 33, 37, 53, 57, 65, 68, 98, 101
Stein, O. 72, 78, 87
Stein, Y. 72, 78, 87
Steinbeck, A. W. 3
Stemerman, M. B. 106
Stetten, M. R. 87, 110
Still, W. J. S. 87, 90, 96
Stockinger, L. 60
Stoebner, P. 43, 44
Stoeckenius, W. 41
Strauss, E. W. 46
Strauss, L. 109
Striker, G. E. 101
Strong, J. P. 90, 108
Strum, J. M. 50
Stump, M. M. 63
Sultan, Y. 106
Sutter, M. C. 19, 76
Švejcar, J. 109
Swadling, J. P. 10

Takada, M. 38, 57, 60
Tanishima, K. 32
Tao, K.-Y. T. 120
Tauchi, H. 66, 67
Taylor, C. B. 90, 93, 118, 131, 132
Taylor, H. B. 38
Taylor, H. E. 98, 102
Taylor, M. G. 13, 106, 107
Telander, R. 63
Telser, A. 110

Tennent, D. M. 123
Tennyson, V. M. 48, 50, 56, 57, 59, 97, 101, 154
Tershakovec, G. A. 152, 154
Terzakis, J. A. 154
Texon, M. 10, 11
Thaemert, J. C. 135, 136
Thoma, R. 46, 61, 63
Thomas, J. 110
Thomas, R. M. 125, 127, 128, 129, 130
Thomas, T. B. 26
Thomas, W. A. 89, 90, 93
Thurau, K. 5
Todd, A. S. 44
Toth, B. 65
Travers, B. 63
Treadwell, C. R. 120
Tristram, G. R. 110
Tromans, W. J. 111
Ts'ao, C.-H. 24, 38, 43, 52, 76, 98, 106, 118, 137
Tsui, C. Y. 66
Tunbridge, R. E. 110, 111, 112

Udenfriend, S. 87, 102
Uehara, Y. 82, 84, 94
Urschel, C. W. 107

Van Citters, R. L. 74, 77, 144
Vane, J. R. 86, 153
Vastesaeger, M. M. 90, 108
Vegge, T. 57, 60
Veress, B. 52, 130, 131, 132
Verity, M. A. 133, 135, 138
Vilar, O. 103
Vitums, A. 24, 37, 38, 60, 68
Vizi, E. S. 81
Vogel, F. S. 65
Vogler, N. J. 102
Voigt, J. 124
Vost, A. 131
Voth, D. 20, 76
Vyslonzil, E. 60

Author index 201

Wagenvoort, C. A. 91
Wagner, B. M. 144
Walford, R. L. 120
Walker, F. W. 6, 91
Walters, V. 20
Wang, H.-P. 107
Warren, B. A. 44, 63, 106, 120
Wasserman, F. 102
Wasserman, K. 27, 53
Watanabe, S. 32
Watson, M. L. 55
Watson, R. F. 97
Watts, H. F. 129, 130, 131, 132
Watts, S. H. 93
Wearn, J. T. 30
Webb, R. L. 19, 69, 74, 75, 76, 85, 148
Webber, W. A. 57, 59
Webster, M. W., Jr 108
Weibel, E. R. 43, 87
Weinstein, M. C. 44
Weis-Fogh, T. 110
Weiss, E. 103
Weiss, J. M. 44
Weiss, L. 90, 91
Weissman, G. 52
Weller, R. O. 108
Werthessen, N. T. 63
Westcott, R. N. 4
Wexler, B. C. 109
Whereat, A. F. 90, 108, 125
Whiffen, J. D. 41
Whipple, A. O. 28
White, J. F. 64
Wicke, H. J. 13, 86
Widmann, J.-J. 26, 27, 50
Wiedeman, M. P. 6, 9, 14, 17
Wiederhielm, C. A. 4, 14, 102, 103, 105
Wiener, E. 94
Wiener, J. 33, 48, 54, 101, 154
Williams, G. 102, 120
Williams, G. M. 61, 63
Williams, R. G. 14, 26, 28, 31, 64, 133, 147
Williams, T. W. 125, 128, 129, 130, 131
Williamson, J. R. 37, 102
Willmer, E. N. 31, 39
Willoughy, D. A. 153
Wilson, R. B. 46, 65
Winternitz, M. C. 125, 127, 128, 129, 130
Wislocki, G. B. 26
Wisse, E. 26, 27, 55, 56, 57, 59
Wissig, S. L. 44, 53, 86
Wissler, R. W. 81, 84, 90, 155
Wittenberg, B. A. 71, 72
Wittenberg, J. B. 71, 72
Woerner, C. A. 52, 127, 128
Wolf, P. L. 64
Wolff, J. 35, 98
Wolfram, C. G. 27, 53
Wolinsky, H. 72, 90, 102, 104, 105, 109, 115, 119, 120, 125, 142
Woo, C.-Y. 20, 86, 136
Wood, E. H. 115
Wood, R. L. 55
Wood, W. B. 150, 152
Woodard, W. C. 31, 48, 63
Woodbury, J. W. 14
Woodruff, C. E. 127, 128
Worthington, W. C. 24
Wortman, B. 89, 90
Wright, I. 118, 123
Wyllie, J. C. 91

Yamada, E. 48
Yoffey, J. M. 2, 3, 130, 131, 153

Zanchetti, A. 5, 139
Zanetti, M. E. 123
Zeigel, R. F. 44, 86
Zelander, T. 32, 57, 60, 74, 80, 82
Zellweger, J. P. 123, 126
Zemplenyi, T. 89, 90, 108
Zimmerman, B. G. 137, 139
Zook, B. C. 108
Zschiesche, L. J. 30
Zuckner, J. 120, 121
Zugibe, F. T. 90
Zumbo, A. 90, 93, 108
Zumbo, O. 90, 93
Zweifach, B. W. 8, 10, 14, 17, 32, 43, 68, 69, 71, 74, 148, 153

SUBJECT INDEX

Acanthosomes 53
Acardiac embryo 120
Accessary hearts 2
Acetylcholine 86, 139
Acid phosphatase 38, 52-3, 96
Actin 81
'Activated' smooth muscle cells 87
Actomyosin 81, 85
  in endothelium 154-5
Adaptation, low temperature, 21
Adenohypophysis 24
'Adenoidal' venules 99
ADP 40
Adrenergic innervation 7, 24, 26, 85, 133, 135, 137
Adrenergic vesicles 133
Adventitial cell 68
Aging of vessels 38, 52, 66, 78, 88, 102, 105, 126
  and arterial calcification 121, 123
  and collagen 106-7
  and elastic tissue 120-1
  and endothelium 44
  and mucopolysaccharides 109
  and physical properties 106
  and vasa vasorum 129
Albumin 27, 60
Alcian blue 32
Alkaline phosphatase 53, 71
Allergic inflammation 154
Allograft 61, 65
  aortic 63
  heart 64
  kidney 48, 64
  skin 48
'All-or-none' flow 77
Amino acids, aging elastin 121
Amphibia 3, 70-1
Anaesthetics 60-1
Anchoring, endothelial 43, 98
Aneurysm 109, 123, 156

aortic 123
berry 123
Angiitis, infectious 129
Angio-endotheliosis 38
Angiotensin 86
Anisotropy
  of collagen 102-3
  of stretched elastica 112
Anoxia 125
'Anoxic enzyme failure' 126
Anoxic necrosis 66
Antibody
  circulating 39
  to endothelium 64
  to platelets 38
  to reticulin 98
Aorta 8, 35, 61, 63, 84, 90-1, 98, 141
  in acardiac embryo 120
  aging in 66, 105, 107, 122
  aneurysms of 123
  of bovidae 128
  canine 13, 128, 130
  as capacity chamber 107
  collagen in 104, 105
  in copper deficiency 124
  degenerative changes in 126
  diameter changes of 144
  development of 86-7, 104, 125
  diffusion in 125-6
  endothelium of 38, 66, 131
  human 13, 77, 107, 115, 128
  intima 77, 105, 130
  'lamellar unit' in 120
  lathyrtic 142
  lipoproteins and 129
  lumen of 144
  lymphatics of 130-1
  mechanical senescence 121
  media of 125-6, 138
  nerves of 133, 138

204 Subject index

Aorta (*cont.*)
  patchy staining trypan blue 129
  permeability: and age 66; of intima 131
  of rabbit 13, 52, 77
  of rat 52, 77, 115
  spontaneous rupture 124
  of swine 77
  syphilis and 123, 132
  vasa vasorum 125, 128, 130
  'Windkessel' function of 13
Aortic arch baroreceptors 140
Aortic bifurcation 11
Aortic blood flow 13
Aortic grafts, prosthetic 63
Aortic valve incompetence 13
Arbovirus 66
Arctic animals 20 *et seq.*
Area
  cross sectional, of vessels 5, 8
  endothelial: aging aorta 66; occupied by fenestrae 57
  exchange, of capillaries 8
  surface, human capillaries 8
Argyria 101
Argyrophil fibres 102–3
Artefacts 55
Arterialization 91
Arterial system, pressure–volume configuration 5
Arterial–venous relationships, particular 20
Arteriole 15, 35, 63, 74, 89, 145
  and axon reflex 140
  and blood pressure 145
  collagen in 105
  diameters 145
  and flare of triple response 140, 152
  in inflammation 152
  innervation of 136, 145
  spiral, endometrium 136
  terminal 74
  vasomotor activity 145
Arterio-portal anastomosis 23–4
Arterio-venous anastomosis (AVA) 28, 76–7, 140, 145–6
  all-or-one activity 119, 146
  in control of heat loss 146
  innervation of 137
Artery 63
  aging of 106, 123
  aneurysms of 123–4
  of arterio-venous anastomosis 117

avian 116
blood flow in 13
calcification with age 123
carotid 115
cerebral 77, 118
coronary 77, 98, 130–1, 137–9
development of 86
dilation of 109
elastic tissue in 117
endothelium 154
innervation 136
intimal plaques 109
intimal smooth muscle 77
leg 123
length vs. diameter changes 104
lymphatics of 130
media of 138
mesometrial 117
muscular 74, 144; collagen in 105, 144; contraction of 116; elastic tissue 116, 144; innervation of 144; nutrition of wall 116
milking through wall of 131
myo-elastic cushions of 118
nerves in 138–9
ovarian 117
parasympathetic nerves of 139
pulmonary 91, 115; aging 120; baroreceptors 140; elastic tissue 120; innervation 137; vasa vasorum 125
renal 85
small 5
splenic 28
umbilical 77, 117
uterine 117
vasa vasorum 125
wall 13; distensibility 103–4; infection of 129; nutrition of 116; tension–length curve 105
Arthus reaction 39
Atavistic property of vessels 20
Atheroma 13, 96, 106, 108
Atherosclerosis 2, 44, 52, 87, 89–90, 93, 107, 109
  aeteology 10, 11, 108–9, 123, 132
  complications 130
  coronary 123
  early changes in 123
  fibrosis in 123
  vasa vasorum and 129–30
ATP 40, 81
ATPase activity, 40, 85

Subject index 205

Atropine 139
　cutaneous reaction to 153
Autofluorescence 102
Autografts 28
Autonomic innervation 23, 86, 136
Autonomic nerves
　fine structure 133, 135–6
　noradrenaline 135
　vesicles of 133, 135
Autoradiography 86–7
Autoregulation 5, 86
'Autoregulatory escape' 139
Avian vessels 83, 91
　innervation 137
Axial stream 9–11, 15
Axon reflex 140, 153

Baroreceptors 132, 138, 140
Barr body 61
Basement membrane 41, 59, 68, 70, 71, 83, 92, 97–8
　and aging 102
　antigens of 102–3
　collagen in 97
　diffusion barrier 99
　elastic tissue relationship 82
　endothelial 42–3
　immunohistology of 97
　multiple 65
　and platelets 106
　and podocytes 101–2
　renewal 102
　and reticulin 102–3
　thickening 102
　triple layered 97–8, 101
　and vasomotor nerve 133
Bat 19
Bat's wing 6, 75–6
Bernard, Claude 1
Birthmark 65
Blood 102
　flow 10, 15 et seq., 77
　fluid mechanics of 13–14
　frictional energy loss 3
Blood platelet 43, 53, 106
　and Arthus reaction 39
　collagen interaction 106
　degranulation 106
　endothelial trophic action 38–9
Blood pressure 3, 4, 50, 63
　and arterioles 145
Blood stream 35
Blood–tissue contact 29

Blood–tissue exchange 101, 125, 148
Blood viscosity 9
Blood volume 3, 23
　in exchange vessels 8–9
　percentage distribution in vessels 6
Body heat, conservation 20
Bone marrow 63
　derived cells 63
Boundary layer 10, 11, 106
Bovine pulmonary vein 118
Bradykinin 61, 86, 152–4
Brain 54
　capillaries, enhanced permeability 53
Branch elastica 115
Branchial vessels 2, 3
Branching of vessels 9, 11
Brownian motion 51
Budgerigar 45, 116
Bulbus arteriosus 115
Bulk flow 54
Buttons of attachment 33

Cadmium 49
Calcification 121
Calcium 78
Campbell de Morgan spots 65
Capacitance vessel 20
Capillariomotor reflex 137
Capillary 2, 57, 69, 75, 93, 99
　accompanying fibres 136
　arterial 29
　in atherosclerotic intima 130
　average length 17
　in bat's wing 17
　bradykinin, action of 153
　contractility 70–1
　definition of 14 et seq.
　densities 8
　endothelium 136
　histamine, action on 153
　growth of 64
　innervation 136
　in vitro 64
　of muscle 17
　permeability 54
　regression of 64
　sensation from 140
　total numbers of 8
　transit time 17
　volume in tissues 8
Carbon dioxide 4, 5, 61
Cardiac cycle 13, 14, 104
Cardiac muscle 91

Subject index

Cardiac output 23
Cardiac systole 129
Caribou 21
Carotid artery 115
Carotid sinus 115, 140
Catecholamine 7, 86, 138–9
  blood 138–9
  fluorescence 133–5, 137
  storage 138–9
Cattle 76
Caveolae 48, 53, 69, 78, 82, 92
Cell degeneration 53
Cell emigration 49
Cell modulation 69, 88
Cell process 68, 92
Cell surface 33
Cellular adhesion 32, 42
Cement substance 32
Centriole 80
Cephalopod 2
Cerebral microcirculation 17
Cerebral oedema 38
Cerebral vessels 33–4, 77
  in hypertension 154
Chains of vesicles across endothelium 55
Channels in endothelium 55
Chemoreceptors 138
Chemotaxis 152
Chick 105
Cholesterol 108, 131
Cholesterol ester 108
Cholinergic innervation 26, 85
Cholinergic vesicles 135
Cholinergic vasodilatation 139
Choriocapillaris 56–7
Chorionic villi 29
Choroid plexus 56, 60, 101
Choroid rete 38, 71–2
Chylomicron 27, 47, 59, 98
Ciliary body 56, 60, 101, 154
Cilium 80, 86
Cinematography
  high speed 11
  time lapse 47
Circle of Willis 123
Circulation
  closed 3
  design faults 3
  in fish 3
  invertebrate 1
  Ohm's law of 4
Clark, E. R. 147

Clearing factor lipase 48
Closing diaphragm 56–7, 59
Coelenterates 1
Cold feet 21
Collagen 86–7, 90, 92, 97, 102 *et seq.*, 110–11, 114–15, 121
  in adventitia 103, 142
  and aging 106–7
  of arterial intima 106
  chemical composition 103
  content of vessels 104
  fine structure of 103
  in lathyrism 142
  microscopy of 103
  muscular arteries 144
  platelet interactions 106
  and properties of vessels 119
  in veins 156
Collagenase 97
Collateral circulation 63
Colloid
  osmotic pressures 148
  removal from bloodstream 26
Colloidal carbon 50, 99
Colloidal tracers 49
Connective tissue 98
Contact inhibition 64, 70
Contractile activity 68
Copper deficiency 124
Coronary artery 29
Coronary atherosclerosis 123
Coronary circulation 129
Coronary evolution 3
Coronary pharmacology 3
Coronary vessels 3, 5
Counter-current system 20, 21, 29, 60, 72
Cow 128
Cremaster 140
Cross-sectional areas, of vessels 17, 145, 156
Cyclostome 24
Cytopempsis 50
Cytoplasmic filaments 69, 80–1
Cytoplasmic microtubules 69
Cytoplasmic ruffles 94

Damage to vessels 63
Death 65
Deep veins 18
Definitions of vessels 14
Desmosine 110
  in copper deficiency 124

Subject index

Desmosome 34–5
Diabetes mellitus 71
Diameter of vessel 14, 89
Diffusion 66, 99
Dilation, vascular 140
Dog 25, 29, 86, 126–8
Doppler technique 13
Double circulation 4
Double ligation 87, 93
Ductus arteriosus, obliteration 118

Ear chamber, rabbit 15–16, 28, 41, 47, 52, 69, 75, 147, 151
Earthworm 97
Elasmobranch 3
Elastase 112–13
   and atherosclerosis 123
Elastic artery 72, 78, 80, 82, 104, 115
   avian 115
   elastic networks of 121
   elastin content 115
   smooth muscle of 119, 142
   strength of wall 141
   wall as two-phase system 119
Elastic laminae 29, 77, 82, 93, 141–2
   fenestrated 113, 115–16
   fibrous splints on 107
   and interlaminar tracts 82
   medio-adventitial margin 116
Elastic tissue 110 *et seq.*
   age changes 120–1
   amorphous areas 114
   and aneurysms 123
   in arteries 144–5
   autofluorescence 110
   basic filaments of 111
   breakdown and removal 123
   calcification 123
   chemistry 112, 113, 121, 123
   collagen fibres in 114
   in copper deficiency 123–4
   development: effect of tension 119–20; *in vitro* 87, 119–20
   distribution of in arteries 117
   elastase action on 113
   embryonic development 120
   fibres 86–7, 92, 103, 112, 114
   fine structure 113
   function in vessels 118–19
   and growth 120
   histochemistry of 113
   histology 112
   hypertension and 120
   inflammation on 123–4
   matrix material 112
   microfibrils 106
   mineralization 121, 123
   peripheral fibrils of 114
   physical properties 118–19, 121
   second protein in 112, 121
   smooth muscle relations 72
   and strength of wall 123–4
   stretching and 112
   of veins 118, 156
   Young's modulus 118–19
Elasticity 118
Elastin 90
   age changes 120–1
   content, arteries 115
   properties 110, 112
   turnover 120
Elastin-like material 87
Elastogenesis 110, 114
Elastomer 110–11
Electron micrograph interpretation 55
Electrotonic coupling 84
Elephant 18
Embryonic vessels 46, 52
Endocrine glands 56
Endo-endothelial coat 32–3
Endoplasmic reticulum 44, 46, 48, 65–6, 69, 78, 86–7, 92, 94
Endothelium 31 *et seq.*, 125
   absence in spleen 28–9
   acid phosphatase in 38
   actomyosin staining 154
   age changes in 44, 46, 66
   alkaline phosphatase in 53, 71
   in allografts 64
   aortic 46
   as barrier 101
   basement membrane of 97–9
   caveolae 48, 50, 61
   contact inhibition of 70
   contraction of 61, 154–5
   denudation of 106
   endoplasmic reticulum 46
   energy demands 48
   fenestrae 56, 60
   flaps of 35, 38, 57
   freeze-etched 56
   fronded cytoplasm 38
   gaps in 27, 38, 61, 99, 154
   Golgi apparatus in 44
   high 155
   hyperplasia of 48

Endothelium (*cont*.)
  immature 46
  in inflammation 150
  in vitro 31, 53
  in vivo 31
  junctions, intercellular 31–3, 35, 37, 53–7, 60–1, 154–5
  lipid inclusions 66
  mitosis in 61, 63
  microtubules in 56
  microvilli 37
  model of 51
  myofibril-like tracts in 71, 154
  necrosis of 66
  nucleus of 61, 66–7
  pericyte relation 69–70
  permeability of 53, 59–61, 131
  physical considerations 39
  plasma membrane 41–2
  pores of 53
  processes of 37, 83
  proliferation of 63–4
  in prosthetic grafts 63
  secretion 33, 43, 103
  sprouts of 47, 98
  swelling of 40
  thinnest regions in 57
  tumours of 65
  vacuoles in 59
  vesicles in 48, 50–1, 55–7, 59–61
En face sectioning 35, 46
Entropy 110
Epinephrine 43
Equine viral arteritis 66
Ergastoplasm 44, 46, 87
Erythrocyte
  goat 15
  parachute deformation 15, 145
Evans blue (T1824) 152–3
Evolution 2
Exanthemata 66
Exchange across walls 64
Exchange vessels 8
Exercise
  and muscle hyperaemia 5
  and venous return 18
External elastic lamina 118
Extracellular fibrils 92, 97
Extracellular fluid 3
Exudation into wall 129

Fenestrae 27, 56–7, 59–60, 65
Ferritin 49, 50, 52, 59

Fibrils, intracellular 92, 94
Fibrin 33, 44, 91
Fibrinogen 44
Fibrinoid 91
Fibrinolytic activity 44
Fibroblast 64, 70, 78, 80, 87, 91–3, 103
  in atherosclerosis 93
Fibroblast–smooth muscle intermediate 91
Fibro-elastic plaque 90
Fibro-elastic reaction 89
Fibro-elastic tissue 92, 105, 115, 146
Fibromuscular laminae 141
Fibrosis 108, 123
Filaments, in collagen and elastica 111
Fine fibrils
  of connective tissue 106, 113
  of endothelium 154
Fish 38, 71
Flare, of triple response 154
Flow
  instability 11
  rates of 5
  turbulent 141
  velocity of 13, 17, 141
Fluid mechanics 13–14
Fluid uptake by vessels 60
Foam cell 90, 96, 108
Foetal liver 26
Foetal pulmonary artery 115
Foetal trophoblast 29
Fowl 24
Freeze-etch 37, 56–7
Frictional shear 9
Frog 101, 105
Frozen feet 21
Functional definition of vessels 14–15

Gas-bubble contact angle 41
Gas gland 72
Giraffe 18, 102, 115
Glass models of vessels 11
Glomerular filtrate 5, 59
Glomerulus 56–7, 59, 69, 101
Glycerination 81, 85
Glycerol–acetic acid esters 61
Glycocalyx 33
Goat erythrocyte 15
Gold-sol 60
Golgi apparatus 43–4, 66, 69, 78, 80, 86–7, 92, 94
Granulation tissue 47, 63–4, 93, 98–9
Granules, autonomic 134–5

Subject index 209

Granulocyte 64
Gravity 76
Great Northern Diver 76
Ground substance 92, 97, 99
Guinea pig 38, 49, 87, 106–7
Gut 56

H-substance 140, 153
Haemangioendothelioma 65
Haemangioma 65
Haematopoesis 26
Haemodynamics 13, 107, 129, 144
Haemoglobin 50, 59
Haemolymph 1
Haemorrhage 26, 65, 130, 150
Haemostasis 41
Hageman factor 152
Halide ions 32
Harvey, William 4
Healing 70, 90
Heart 3, 18, 29, 102, 107, 130
   peripheral 156
Heparin 32, 41, 48
Hepatic artery 23–4
Hepatic blood flow 23–4
Hepatic circulation 27
Hepatic lymph 28
Hepatic-portal system 23–4
Hepatic sinusoids 23, 27, 59
Highly permeable vessels 61, 101
Hind leg preparation 54, 90
Histamine 49, 61, 85, 153–4
Histiocyte 93
Holes in endothelium 57
Hormonal control of vessels 5, 148
Horse 66
Horseradish peroxidase 49–50, 54, 59, 60
Human 77, 85, 91, 105, 107, 109, 112, 117, 121, 123, 125, 127–8
Hydraulic hindrance 10
Hydraulic models 9–10
Hydrocephalus 60
Hydropic swelling, endothelium 66
Hydrostatic pressure 148
Hydroxyproline 87, 103, 110
Hyperaemia 5
Hyperosmolality 5
Hyperplasia 48
Hypersensitivity 39, 65
Hypertension 50, 52, 54, 63, 90–1, 109, 115, 120–3, 154
Hypertrophy 120

Hypophyseal portal system 38
Hypoxia 5, 25, 64, 126

Immunofluorescence 81
Immunohistochemistry of collagen 103
Impedance, non-uniform 107
Impedance method 14
Incompetent valves, venous 18
Inflammation 18, 37, 41, 52, 61–3, 69, 70, 98, 108, 123, 129, 150, 152–5
Injection of lymphatics 130
Innervation of vessels 133 *et seq.*
Intercellular contacts 83–4
Intercellular spaces, endothelial 47
Interfacial forces 40
Intermittent flow in vessels 13, 129
Internal elastic lamina 83, 116, 117, 118, 123, 144
Intestinal villus 60
Intimal cushion 77
Intimal damage 11
Intimal fibroblasts 93
Intimal proliferation 89
Intimal repair 118
Intimal thickening 93
Intramural tension 129, 141
Intravascular necklace 58
Invertebrates 1, 2
In-vitro observations 48, 64, 94
In-vivo observations 19, 23, 28, 31, 43, 55, 60, 70, 76, 85–6, 94, 102, 133, 139, 144–5, 147–8
In-vivo techniques 147
Ions 78, 99
Ischaemia 129
Isoporous membrane 54

Juxtaglomerular apparatus 89

Kidney 56, 72, 89, 135, 152
Kupffer cell 27–8, 50, 66

Labour 77
Lamb 127, 136–7
Laminar flow 9, 11
Laminated basement membrane 82, 98
Langhans cell 77
Lanthanum 54
Laplace equation 119
Large pores 53, 60
Lathyrism 142
Leaks, vascular 54

210 Subject index

Leukocyte 37, 42, 98, 150, 152–5
  emigration 150, 152–5
  margination 150, 152, 155
Leukotaxine 152
Limb vessels 20
Lipid 46, 80, 90, 96, 109, 113, 123, 131
  infiltration 107 *et seq.*
Lipophage 96
Lipoprotein 59, 129, 131
Liver 23, 54, 56, 65, 69, 136
Lizard 137
Low permeability vessels 101
Lumen, vascular 8, 9, 46, 77, 141, 146
Lymph 155
Lymphatic endothelium 53
Lymphatic system 2, 3
  of vessel walls 130 *et seq.*
Lymphocyte 99, 155
Lymphoid tissue 18, 26, 28, 99, 100, 155
Lysosome 52–3, 80, 94, 96

Macromolecules, permeability to 53
Macrophage 28, 40, 43, 50, 64, 71, 90, 93–6
Macula occludens 33
Magnesium 5
Malpighi 14
Malpighian body 28
Mammary gland 52
Marfan's syndrome 124
Mechanical damage, endothelium 61
Mechanical stimulation, smooth muscle 85
Mechanoreceptor 86
Megakaryocyte 39
Melting point, body fats 21
Membrane fusion 33
Membrane nexus 84, 92
Mesenchymal cells 69, 70–1
Mesothelium 39, 48
Metabolic control of fenestrae 59
Metabolic effects, arterioles 145
Metabolite exchange 9
Metarteriole 15, 74–5, 145
Metastasis 19, 26, 65
Micro-angiogram 107
Microcannulation 14
Microcirculation 147 *et seq.*
  blood flow in 14–15
  function of 147
  leukocytes in 150
  macrophages and 96
  model of 10

network vessels 9
pericytes in 70
permeability changes in 155
smooth muscle in 85
specific patterns of 147–8
Starling's hypothesis in 148
vasa vasorum 128–9
vasomotion in 150
in vessel walls 129
Microfibrils, endothelial 43
Microtubules, endothelial 43
Microvillus 37, 59–60
Mitochondria 48, 69, 78, 87, 94, 138
Mitosis 61, 69, 70
Model, computer 51
Model for capillary permeability 53–4, 60
Monocyte 90, 96
Morphine 153
Morphometry 104
Mouse 20, 28, 36, 49, 50, 66, 106, 107
Mucopolysaccharide 33, 109, 123
Multivesicular body 80
Mural cells, retina 70–1
Mural tension 109
Mural thrombus 44, 130
Muscle capillaries 54
Muscle contraction and venous return 18
Myelinated nerves to vessels 138, 140
Myeloid series 50
Myeloperoxidase 59
Myocardial infarct 123
Myo-elastic cushions 118
Myo-elastic tissue 83, 114
Myofibrils 78, 80–2, 87
Myo-intimal cell 87
Myometrium 88
Myosin 81, 82
Myxini 2, 26

Necks of vesicles, endothelial 55, 57
Necrosis
  in atherosclerotic plaque 130
  of foam cells 108
  tissue 150
  of tunica media 118, 126
Nerves
  of vessel walls 5, 133 *et seq.*, 138–9;
    plexuses of 137–8
  of vessel and triple response 152
Nervous regulation, microcirculation 148

Subject index

Neuromuscular control, veins 75
Neuromuscular synapse, vascular 133, 135–6, 138
Neurotransmitter 133
New vessel formation 46, 64, 70, 98
Non-lipid-soluble materials, transport 53
Norepinephrine (Noradrenaline) 85, 135, 139
Noxious stimuli 48
Nuclear membrane 61, 78
Nuclear swelling, endothelial 69
Nucleolus 69, 77, 92, 94
Nucleus
  of endothelium 61, 66–7
  of fibroblast 92
  of macrophage 94
  of pericyte 69
  of smooth muscle 77–8
Nutrient vessels 68, 125, 145, 148
Nutrition of arterial wall 116

Octopus 136
Oedema 150, 153
Oestrogen 86–7
Oil-water partition coefficient 61
'One-way traffic' across endothelium 42
Oral contraceptives 38
Organelle-rich zones 78
Organization, thrombus 106
Osmium tetroxide, and elastica 112
Oxygen 4, 17, 29, 61, 71–2, 125

Pain, bradykinin causing 153
Pampiniform plexus 22
Pancreas 56
Parabolic flow profile 10–11
Parachute cells 15, 34
Parasympathetic innervation 139–40
Penicillate arteries 28
Perfusion fixation 57, 104, 115, 142
Pericyte 42, 64, 68–72, 75
Permeability 18, 32, 37–9, 50, 53, 101, 129, 155
  increased 153–4
Peroxidase, endogenous 27, 50
Peyer's patch 100, 155
pH, and endothelial growth 64
Phagocytic activity 27, 52, 68, 71, 93
Phagosome 94
Phase effects and pulse obliteration 14
Phases in vesicular transport 50–1
Phasic contractions, veins 19

Phospholipid 108
Pig 87, 104, 126, 132, 139
Pinocytosis 1, 38, 50, 72
Pituitary, portal system 24
Placenta 29
Planimetry 105
Plaque, atherosclerotic 13, 130
Plasma 3, 47
Plasma membrane 48, 50–1, 56, 69, 80, 82, 92, 97
Plasminogen activator 44
Plastic hub 63
Platelets, blood 38
Platyhelminthes 1
Podocyte 101–2
Polarization of endothelium 42
Polarized light 102, 112
Polysaccharide 40, 99, 102, 113
Pores, physiological 55
Portal systems 23
Portal vein 20, 24, 76, 81, 91, 118
Post-capillary venule 17–18, 51, 155
Post-exercise hyperaemia 5
Post-stenotic dilatation 11
Potassium 5
Pre-atheromatous lesion 93
Pre-capillary sphincter 23, 75
Pregnancy, arteries and 117
Pressure–flow relations, 89, 109
Proline 86–7, 103
Prostaglandins 86
Prostheses 38, 63, 120
Protein 44, 49, 53, 60, 82, 86, 103, 111, 131, 150
Protozoa 1
Pulmonary artery 26
Pulmonary circulation 4, 14, 25
Pulmonary cutaneous system 3
Pulmonary hypertension 91, 115
Pulmonary oedema 4
Pulmonary vein 25, 76
Pulsatile flow 13
Pulse wave velocity 106
Purinergic endings 86

Quintuple-layered membranes 33, 84

Rabbit 29, 34, 38–9, 46, 74–5, 77, 87, 89, 99, 101, 106, 109, 118, 126
Radio-opaque injection 24
Rat 20, 22, 26, 35, 41, 45, 49, 50, 62, 66, 73, 79, 84, 86–7, 90, 100–1, 104, 106–7, 120, 122, 126, 131–2, 134, 153

212 Subject index

Reactive endothelium 48
Red blood cell 15, 26, 29, 34
Re-endothelialization, of allograft 61
Renal capillaries, peritubular 24, 160
Renal portal system 24, 137
Renal tubule 59
Renal vessels 57, 63, 72
Repair of endothelial defect 61
Repair process
    in vessel walls 87, 93, 96, 109
    in wound healing 64
Reptilia 4
Resistance vessels 91
Resonance, arterial wall 13
Rete mirabile 35, 60, 72
Reticulin 102
Reticulo-endothelial system 26, 29, 103
Retina, vessels in 66, 70–1, 76, 136
Retrograde flow, aortic 13
Reynold's number 11
Rheology 77
Rhythmic contractions, of vessels 76, 86, 91, 156
Rhythmic forces and elastogenesis 119
Ribosomes 48, 80, 94
Rickettsia 40
Rouget cell 68, 71
Roughness of vessel wall 9
Rounding-up, of cells 40
Rubber bladder–string net model 141–2
Rubella 66

^{35}S studies, endothelium 33
Saccharated iron oxide 60, 99
Scars 109, 120
Schwann cell 133, 135
Sclerosis, arterial 106
Scrotal sac 22
Seagull 21
Seamless vessels 35–6
Secretion, endothelial 43, 46
Seeding of endothelium 63
Selective valving 24–5
Senile angioma 65
Senile atrophy 66
Sensory innervation of vessels 140
Serotonin (5-HT) 49, 61, 86, 153–4
Servetus, Michael 4
Seven-layered attachments 84
Sex chromatin 61
Sex hormone 117
Shared antigens, platelets and endothelium 39

Shear, high rates beneficial 11
Sieve-plate endothelium 27, 59
Silver 31, 69, 101–2, 133
Sinusoids 2, 3, 24, 26–30, 59, 101
Skeletal muscle 85
Skin, responses of vessels 140
Slack collagen fibres 105
Sliding filament hypothesis 81
Slits in vein walls 28
Small pores 53
Smooth muscle
    actomyosin content of cells 85
    of aorta 67, 126
    of arteries 74
    of arterioles 74, 145
    of arterio-venous anastomosis 76–7, 146
    in atherosclerosis 89–90
    attachments of 82–4, 119
    autoregulation of 144
    basement membrane 82
    caveolae 48
    cilium 80
    contraction 84–6
    dimensions of 72
    of elastic arteries 72, 115, 119
    endoplasmic reticulum of 78, 86–7
    excitation 84, 86, 138–9
    and fibroblast 88
    fibroblastic activity of 86–7
    fine structure of 77 et seq.
    Golgi apparatus 86–7
    hyperplasia of 90
    in hypertension 90
    hypertrophy of 90
    innervation of 135–6, 144
    intimal 77, 117
    maturation of 78
    multifunctional cell 84
    myofibrils of 81
    necrosis of 66, 91
    orientation of 75
    primitive 75
    relation to Rouget cell 68–9
    in repair of vessel wall 93
    secretion by 89
    single-unit type 86
    in sphincters of vessels 28
    tension generated by 85
    of veins 76, 137, 156
    of venules 75
    vesicles in 80

Subject index 213

Smooth muscle-like contraction, endothelium 154
Smoothing 4, 14, 129
Solid cords, endothelium 46
Span cell 82–3
Sphincter, vascular 23, 28, 74, 136, 145
Sphincter-like bodies 76
Spermatic fascia 22
Spermatozoa 22
Spiral cushions, intimal 77
Spiral valve, of heart 3
Spiralling of arteries 106
Spleen 28–9, 49
Splinting of endothelium 43
Starling's hypothesis 148
Stasis 150
Stereology 87, 115
Steroids 38
Stresses, in vessel walls 103, 105
Striated muscle, of vein walls 20
Sulphonic acid 32
Surface-active agents 40–1
Surface area, 1 ml of blood 8
Surface charge 42
Surface energy 39–40
Surface properties, endothelial 41
Swim bladder 35, 60, 72
Sympathectomy 140
Sympathetic nervous system 5, 136, 139, 144–5
Sympatheticomimetics 29
Syncope 2
Syphilis
  aortitis due to 123
  lymphatic spread in 132
Systemic circulation, pressure maintenances in 4

Teleost 3, 35, 72
Tensile strength 121
Tension, in wall 119
Tension–length curve 105
Testicular artery 22
Testis 22
Thermal injury 154
Thermodynamics 40
Thermoregulation 22
Thoracic duct 130
Thorium dioxide 51, 60, 99
'Thrombo-hemolytic thrombo-cytopenic purpura' 39
Thromboplastic activity 43–4

Thrombosis 41, 44, 64, 66, 130, 150
Thrombus 65, 106, 109
[^{3}H]Thymidine 28, 63
Thyroid 50
Tight junctions 33–4, 38, 136
Time-lapse cinematography 47
Tissue culture 39, 53, 63, 87, 92
Tracer particles 98
Transit time, arteriole–venule 17
Transmitter, neuromuscular 138
Triglyceride 47
Triple response 140, 150, 152
Trophic changes 2
Tropocollagen 103
'Tropoelastin' 114
Trypan blue 129, 153
Tumour 19, 65
Tunica adventitia 91–3, 96, 103–4, 118, 127, 130, 137–8, 141–2
Tunica intima 105, 109, 128, 130, 138
Tunica media 72, 76, 83–4, 89, 91, 93, 105, 109, 126, 128, 130, 137–8, 141
Turbulent flow 11
Two-phase system 119, 121, 124

Ultrafiltrate, plasma 59
Ultrastructure, interpretation 55
Umbilical vessels 20, 63, 136
Unit membrane 33, 43, 56, 82
Uraemia 126
Ureter 89
Urine 102

Vasa vasorum 63, 96, 123, 125 et seq., 131
Vascular conductance 4
Vascular grafts 93
Vascular hyperplasia 63
Vascular occlusion 48
Vascular patterns, of tissues 149
Vascular permeability 48, 50, 99, 148–50
Vascular relationships, in limbs 20
Vascular resistance 4, 9
Vascular tone 4
Vaso-active materials 85
Vasoconstriction 153
Vasodilatation 86, 139
Vasomotion 69, 84, 145, 153
Vasomotor control 21, 139–40
Vasomotor nerves 135
Vasomotor responses, graded 119
Vasopressin 86

Subject index

Veil cell 92
Veins
  accessary hearts of 2, 26
  bursting pressure of 105
  as capacitance vessels 6
  as collecting vessels 17 *et seq.*
  as conduits 155–6
  contractions of 19, 20, 24, 86, 156
  in diving animals 6, 76
  elastic tissue in 118
  general 155–6
  as grafts into arteries 120, 156
  innervation of 137, 140
  in invertebrates 1
  lymphatics in walls of 130
  pulmonary 20, 118
  as pumps 18
  smooth muscle in 76, 137
  striated muscle in 91
  tone changes 6–7
  tunica media of 76
  valveless 18–19
  valves in 18–19, 24, 75
  varicose 18, 109
  vasa vasorum of 125, 128
  vertebral system 19
  and volume-flow alterations 156
  wall, infections of 129
Velocity of blood flow 10
Vena cava 6, 52, 76
Venae commitantes 18
Venous capillary 57, 60
Venous limb of AVA 117, 145–6
Venous plexus 19
Venous return, 7, 18–19
Venous sinuses, vertebral 25
Venous siphon 145–6
Venous system 1, 6

Venules 15, 17–18, 54, 57, 60, 66, 75, 92, 150–2, 154–5
Vesicles
  in endothelium 48, 50–1, 55–7, 59–61
  in smooth muscle 80
Vessel lumen 93
Vessel strip preparation 85
Vessel volume distensibility 106
Vessel wall
  critical thickness 125
  properties modified 119
  stiffness 105–6
  strength 103
Vibrations, arterial 13
Virus and endothelium 40, 65–6
Viscosity 11, 43
Viscous shear 10
Volume of blood in tissues 8
Voluntary muscle 18

Wall tension 120
Wall-to-lumen ratio 91
Wall stresses 124
Water, movement 53–4, 60, 72
'Watershed' effect, aorta 125–6
Wear-and-tear 13
Weibel and Palade body 43–4
Whale 20
'Windkessel' function 13
Wound healing 52, 63–5, 69, 98

^{133}Xe blood flow measurement 18

Young's modulus 118

Zimmerman, K. W. 69
Zonula adhaerens 34, 70
Zonula occludens 33, 54–5

DATE DUE

MAR 1 2 2002

596.0116 C61   96964

CLIFF

BLOOD   VESSELS

College Misericordia Library
Dallas, Pennsylvania 18612